Health in the green economy

Health co-benefits of climate change mitigation

Housing sector

WHO Library Cataloguing-in-Publication Data

Health in the green economy: health co-benefits of climate change mitigation – housing sector.

1.Climate change. 2.Housing - standards. 3.Environmental health. 4.Conservation of natural resources. 5.Risk assessment. I.World Health Organization.

ISBN 978 92 4 150171 2 (NLM classification: WA 795)

© World Health Organization 2011

All rights reserved. Publications of the World Health Organization are available on the WHO web site (www.who.int) or can be purchased from WHO Press, World Health Organization, 20 Avenue Appia, 1211 Geneva 27, Switzerland (tel.: +41 22 791 3264; fax: +41 22 791 4857; e-mail: bookorders@who.int). Requests for permission to reproduce or translate WHO publications – whether for sale or for noncommercial distribution – should be addressed to WHO Press through the WHO web site (http://www.who.int/about/licensing/copyright_form/en/index.html).

The designations employed and the presentation of the material in this publication do not imply the expression of any opinion whatsoever on the part of the World Health Organization concerning the legal status of any country, territory, city or area or of its authorities, or concerning the delimitation of its frontiers or boundaries. Dotted lines on maps represent approximate border lines for which there may not yet be full agreement.

The mention of specific companies or of certain manufacturers' products does not imply that they are endorsed or recommended by the World Health Organization in preference to others of a similar nature that are not mentioned. Errors and omissions excepted, the names of proprietary products are distinguished by initial capital letters.

All reasonable precautions have been taken by the World Health Organization to verify the information contained in this publication. However, the published material is being distributed without warranty of any kind, either expressed or implied. The responsibility for the interpretation and use of the material lies with the reader. In no event shall the World Health Organization be liable for damages arising from its use.

Design by Inís Communication – www.iniscommunication.com

Cover photo: A team of workers affix a passive solar hot water heater to a home in Kuyasa, South Africa, as part of an energy efficiency and housing upgrade project in a low-income area of Cape Town. The project has been credited with helping to reduce a range of housing-related diseases. It is also the first in South Africa to qualify for finance under the Clean Development Mechanism of the United Nations Framework Convention on Climate Change (UNFCCC). See Case Study, Chapter 6. (Photo credit: Nic Bothma/Kuyasa CDM)

Printed in Malta

Acknowledgements

This document draws considerably upon the discussions of an international meeting of experts on *Housing, health and climate change*, hosted by the World Health Organization in October 2010. Participants who contributed to early drafts and reviewed the final document are acknowledged here:

Humaid Abdulla Al Marzouqi, Abu Dhabi Urban Planning Council, United Arab Emirates

John M. Balbus, National Institute of Environmental Health Sciences, United States of America

Carlos Barcelo Perez, Instituto Nacional de Higiene, Cuba

Matthias Braubach, WHO European Centre for Environment and Health, Germany

Charlotte Bringer-Guérin, Ministry of Health, France

Emmanuel Briand, Ministry of Health, France

Ellen Winters Daltrop, Housing Policy Section, UN-HABITAT, Kenya

Ulla Haverinen-Shaughnessy, National Institute for Health and Welfare, Finland

Philippa Howden-Chapman, New Zealand Centre for Sustainable Cities and He Kainga Oranga Housing and Health Research Programme, University of Otago, New Zealand

David Jacobs, National Center for Healthy Housing, United States of America

He Jianqing, National Engineering Research Centre for Human Settlements, China

Amir Johri, WHO Centre for Environmental Health Activities, Jordan

Laura Kolb, Office of Radiation and Indoor Air, Environmental Protection Agency, United States of America

Cornis Lugt, Division of Technology, Industry and Economics, United Nations Environment Programme, France

Rebecca Morley, National Center for Healthy Housing, United States of America

Nisha Naicker, Environment and Health Research Unit, Medical Research Council, South Africa

Toshitaka Nakahara, Medical Department, Kyoto University, Japan

Guillerme Netto, Ministry of Health, Brazil

Natheer N. Abu Obeid, Faculty of Architecture and Design, Jordan University of Science and Technology, Jordan

David Ormandy, WHO Collaborating Centre for Housing Standards and Health, Institute of Health, University of Warwick, England

Karin Otzelberger, WHO Collaborating Centre for Housing and Health, Germany

Annette Rebmann, WHO Collaborating Centre for Housing and Health, Germany

Abdul Saboor, Construction and Urban Development, International Labor Organization, Switzerland

Marinus Verweij, Netherlands Organization for Applied Scientific Research (TNO), Built Environment and Geosciences, Netherlands

Claudia Weigert, Environmental Health Division, Directorate-General of Health, Portugal

Edmundo Werna, Sectoral Activities, International Labour Organization, Switzerland

Paul Wilkinson, Department of Social and Environmental Health Research London School of Hygiene and Tropical Medicine, England

Sergii Yampolskyi, Environment, Housing and Land Management Division, United Nations Economic Commission for Europe, Switzerland

Hina Zia, Centre for Research on Sustainable Building Science, The Energy and Resources Institute (TERI), India

External review was also provided by:

Claudio Acioly, Housing Policy Section, UN-HABITAT, Kenya

Rajiv Bhatia, Occupational and Environmental Health, San Francisco Department of Public Health, United States of America

Genon Jensen, Health and Environment Alliance (HEAL), Belgium

Mili Majumdar, Sustainable Habitat Division, The Energy and Resources Institute (TERI), India

Angela Mathee, Environment and Health Research Unit, Medical Research Council, South Africa

Jean-Luc Salagnac, French Scientific and Technical Centre for Building (CSTB), France

Content for this report was developed by:

Project co-ordinator:
Carlos Dora, Department of Public Health and Environment, World Health Organization

Lead author:
Nathalie Röbbel, Department of Public Health and Environment, World Health Organization

Editor:
Elaine Ruth Fletcher, Department of Public Health and Environment, World Health Organization

Administrative support:
Pablo Perenzin, Saydy Karbaj, Eileen Tawffik and Terri Mealiff

Graphic design:
Aaron Andrade, Inís Communication

The following WHO personnel also provided review and contributions in their field of expertise:

Heather Adair-Rohani, Department of Public Health and Environment, World Health Organization

Ahmad Basel Al-Yousfi, WHO Centre for Environmental Health Activities, Jordan

Robert Bos, Department of Public Health and Environment, World Health Organization

Diarmid Campbell-Lendrum, Department of Public Health and Environment, World Health Organization

Yves Chartier, Department of Public Health and Environment, World Health Organization, Switzerland

Chee Keong Chew, Department of Public Health and Environment, World Health Organization, Switzerland

Ruth Etzel, Department of Public Health and Environment, World Health Organization

Ivan Ivanov, Department of Public Health and Environment, World Health Organization

Olaf Horstick, UNDP/UNICEF/World Bank/WHO Special Programme on Research and Training in Tropical Diseases (TDR)

Axel Kroeger, UNDP/UNICEF/World Bank/WHO Special Programme on Research and Training in Tropical Diseases (TDR)

Mazen Malkawi, WHO Centre for Environmental Health Activities, Jordan

Marina Maiero, Department of Public Health and Environment, World Health Organization

Bettina Menne, WHO European Centre for Environment and Health, Italy

Suzanna Martinez Schmickrath, Health Security and Environment Cluster, World Health Organization

Carolyn Vickers, Department of Public Health and Environment, World Health Organization

Elena Villalobos, Department of Public Health and Environment, World Health Organization

Susan Wilburn, Department of Public Health and Environment, World Health Organization

Tanja Wolf, WHO European Centre for Environment and Health, Italy

WHO Regional Focal Points (Health Environment and Sustainable Development):

Lucien A Manga, WHO Regional Office for Africa (AFRO)

Luiz AC Galvao, WHO Regional Office for the Americas (AMRO)

Mohamed Aideed Elmi, WHO Regional Office for the Eastern Mediterranean (EMRO)

Jai Narain, WHO Regional Office for South-East Asia (SEARO)

Srdan Matic, WHO Regional Office for Europe (EURO)

Michal Krzyzanowski, WHO Regional Office for Europe (EURO)

Hisashi Ogawa, WHO Regional Office for the Western Pacific (WPRO)

Development of this report was supported financially by the Government of France and the United States Environmental Protection Agency (US EPA)

Foreword

Evaluation of the health impacts of climate mitigation strategies is critical to informed decisions that will attain the greatest combined gain for health, well-being and sustainable development.

This report considers the scientific evidence regarding possible health gains and, where relevant, health risks of climate change mitigation measures in the residential housing sector. The report is one in a *Health in the Green Economy* series led by WHO's Department of Public Health and Environment. Other reports in the series focus on transport, household energy in developing countries, agriculture and health care facilities.

(Photo: Nic Bothma/ Kuyasa CDM)

The focus of analysis is mitigation measures discussed in the *Fourth Assessment Report*[i] of the Intergovernmental Panel on Climate Change, which represents the UN system's most broad-based scientific assessment of mitigation options. The aim is thus to provide health-oriented review of mitigation strategies around which broad scientific consensus already exists as to impact and feasibility.

The report documents how certain mitigation options can yield substantial co-benefits to health. Some choices, however, may be better than others in terms of health impacts, or reducing health risks. New and sometimes overlooked opportunities are also examined where health gains and sustainability objectives can be mutually reinforcing.

These findings have twofold relevance.

For the health community, they represent a major opportunity to promote "primary prevention" by informing policy-makers and the public about how better health can be obtained from economic investments in housing.

Also, evaluation of health impacts touches the core of a debate that has stalled climate change negotiations – the debate about who 'gains' and who might 'lose'. Looking at health co-benefits creates a different paradigm, one that is 'win-win' for most people, and for the planet.

In fact, there is good evidence that many climate mitigation strategies can yield both immediate and more sustained global public health benefits – in rich and poor societies, temperate or tropical, urban and rural.

Often these health benefits can be derived at comparatively low cost – and at almost no cost to resource-strapped health services – but rather through more careful, strategic, and health-focused development investments.

---

[i] Metz, B et al. eds. *Climate Change 2007: Mitigation of Climate Change. Contribution of Working Group III to the Fourth Assessment Report of the Intergovernmental Panel on Climate Change.* Cambridge & New York, Cambridge University Press, 2007.

In terms of housing, for instance, health benefits may be derived when mitigation policies also: improve indoor air quality; reduce exposure to heat waves and extreme cold; prevent vector and pest infestations; prevent home injuries; improve safe drinking-water and sanitation access; avoid use of toxic and hazardous construction materials; reduce vulnerability to floods, mud slides and natural disasters; and support slum redevelopment and better environmental design of transport, energy and utility infrastructure in fast-growing developing cities.

For some key measures, there is quantifiable evidence of economic savings in health care costs. For instance, investments in home insulation have reduced health care costs of chronic respiratory disease in some settings, justifying investments made in large-scale housing improvement programmes. More such cost-benefit evaluation would likely make the case for action even more compelling.

This report identifies key "co-benefits" themes and possibilities. More evidence, however, is needed about health impacts of specific interventions. Also, development policies, subject to multiple political and economic forces, can be difficult to implement – even with the powerful logic of health. Still, the challenges of action are no excuse for inaction.

Addressing pressing public health priorities through mitigation strategies is clearly better than not doing so. Such strategies, once understood, are likely to receive broad public and political support. By identifying what health gains are expected and how health may benefit, we also contribute to the dialogue about how different models of production and consumption impact on the epidemic of noncommunicable disease, on resurgence or emergence of new disease epidemics and on the stubborn intransigence of certain diseases of poverty.

Let us also acknowledge that the ultimate goal of climate change mitigation is better human health and well-being. So why not make it central to our strategies as well?

By refocusing the debate around health, with responsible use of evidence, we translate abstract climate concepts into impacts on diseases and issues that people know and care about. This series makes the fundamental case that investments in climate change mitigation can produce better health, at lower cost, if informed decisions are made. This advances the goal, as articulated in WHO's 1948 Constitution, to promote "the highest attainable standard of health" as "one of the fundamental rights of every human being without distinction of race, religion, political belief, economic or social condition." This series seeks to outline such important opportunities.

Dr Maria Neira
Director of Public Health and Environment
World Health Organization

Contents

EXECUTIVE SUMMARY 1

    Key messages 1

    Background and rationale 3

    Scope and methods 4

    Summary of findings 4

INTRODUCTION 9

    Background and rationale 9

    Scope and methods 10

1 OVERVIEW OF HOUSING AND CLIMATE/ENVIRONMENT LINKAGES 15

    1.1 How housing contributes to climate change 15

    1.2 Trends in developed versus developing countries 16

    1.3 Housing density and urban design as factors in GHG emissions 18

    1.4 Slums and their environmental/climate change impacts 20

    1.5 Regional climate-related impacts on housing environments 21

    References 26

2 REVIEW OF HOUSING AND HEALTH RISKS 29

    2.1 A framework for understanding health risks in housing 29

    2.2 Environmental health risks 30

    2.3 Diseases and injuries 38

    References 41

3 EVALUATING HEALTH CO-BENEFITS AND RISKS OF IPCC-REVIEWED MITIGATION STRATEGIES 45

    3.1 Methods of analysis 45

    3.2 Scope of mitigation issues considered 45

    3.3 Limitations of the analysis 46

    3.4 Thermal envelope 47

    3.5 Heating systems 52

    3.6 Cooling loads 56

    3.7 Whole-building heating, ventilation and air conditioning systems (HVAC) and space/unit air conditioners 61

    3.8 Passive and photovoltaic solar energy 65

 3.9 Lighting and day lighting 67

 3.10 Household appliances and electronics. 69

 References 74

4 GAP ANALYSIS: OPTIMIZING HEALTH BENEFITS AND CORRECTING RISKS OF MITIGATION STRATEGIES 79

 4.1 Health co-benefits and risks of IPCC-reviewed mitigation measures 79

 4.2 Neglected co-benefit opportunities: healthy urban design 80

 4.3 Addressing housing and health inequities with mitigation strategies 85

 4.4 Occupational health – risks and exposures to construction and building renovation workers . 88

 4.5 Behavioural change: factors that promote or confound strategies 89

 References 91

5 TOOLS TO ASSESS, PLAN AND FINANCE HEALTHY AND CLIMATE-FRIENDLY HOUSING 93

 5.1 Assessment methods (HIA, SEA, EIA) 93

 5.2 Intervention studies 94

 5.3 Indicator systems 95

 5.4 Regulatory frameworks, including building and planning codes 96

 5.5 Tools for financing interventions 98

 References 101

6 CASE STUDIES OF GOOD PRACTICE 103

 6.1 New Zealand's *Housing, insulation and health* study 103

 6.2 Lighting a Billion Lives 104

 6.3 Low-cost urban housing energy upgrade project in Cape Town, South Africa 106

 6.4 Low-carbon housing measures and vector-borne disease control 108

 References 109

7 CONCLUSIONS AND RECOMMENDATIONS 111

 7.1 Largest health co-benefit opportunities 111

 7.2 Health risks to be avoided 113

 7.3 Gaps in current mitigation analysis 114

 7.4 Implementing win-win health, housing and climate change strategies 117

 7.5 Regulatory frameworks 118

 7.6 Climate finance for health 119

 7.7 Building community capacity 120

Executive summary

Key messages

Health co-benefits of housing-related climate change mitigation

- **Investments in climate-friendly and energy-efficient housing** can significantly reduce transmission of infectious diseases as well as help prevent many noncommunicable diseases.

- **Noncommunicable diseases.** Cardiovascular disease, strokes, injuries, asthma and other respiratory diseases, can be prevented through low-energy and climate friendly housing measures that:
  - reduce exposure to extreme heat and cold;
  - reduce exposure to mould and dampness;
  - improve indoor air quality through better natural ventilation;
  - provide for safer, more energy-efficient home heating and appliances.

- **Communicable diseases.** Transmission of airborne and waterborne infectious diseases and of certain vector-borne diseases can be reduced through low-energy and climate-friendly housing designs that:
  - increase natural ventilation to lower risks of airborne infection transmission, including tuberculosis (TB);
  - limit vector and pest infestations through measures such as sealing of cracks and window screening;
  - improve access to safe drinking-water and improved sanitation in planning and siting.

- **Mental health and sense of well-being.** Many housing improvements can improve both.

- **Health can be an economic driver of housing investments.** While climate gains may be mostly reaped in the future, many of the health gains are immediate and quantifiable. These include savings to households, health systems and economies in terms of reduced illness, fewer medical visits and sick days off work and school.

- **Health equity can be a major co-benefit of better planned, more energy-efficient housing.** Cleaner cookstoves can help save the lives of nearly 2 million people annually who die from respiratory diseases related to household air pollution, including chronic pulmonary diseases and pneumonia.[i] Better urban planning can reduce macro-energy costs of housing while improving access to securely-sited, safely-constructed homes with utility and transport services, as well as green spaces for physical activity and positive social interaction. These in turn can help prevent illnesses related to heat waves, unhealthy housing and poor urban environments that affect the poor.

Win-win strategies for greener housing design – and how they benefit health

- **Improved design for natural ventilation** can increase fresh-air exchanges and thus indoor air quality, reducing:
  - risks of airborne infection transmission, including TB;
  - buildup of, and exposure to, toxic indoor air pollutants from interior design materials, furnaces (e.g. carbon monoxide) and naturally occurring radiation (radon);
  - mould/dampness that are risk factors for allergies and asthmas;
  - indoor air pollution risks from air conditioning systems, including buildup of micro-organisms (e.g. *legionella*).

- **In areas that are heavily polluted outdoors, air filters can help reduce indoor air pollution.**

>>

---

[i] See the companion *Health and Green Economy* report: *Co-benefits to Health of Climate Change Mitigation: The Household Energy Sector in Developing Countries* (Adair-Rohani H, Bruce N, 2011).

- **House screening of doors and windows can be an effective, low-energy vector-control tool.** Screening has been neglected as reliance on bednets, chemical insecticides and air conditioning increased. In light of the greater awareness of natural ventilation's health benefits, and the energy demands of air-conditioning, house screening deserves renewed attention as a complementary vector-control measure.
- **Improved thermal insulation can reduce risks of respiratory and infectious diseases related to cold, damp, and mould exposures, as well as illness from pests.** Good ventilation is, however, critical to ensuring health gains from energy-efficient and weather-tight housing.
- **Greener construction materials can reduce health risks from toxic chemical exposures.** The health risks of asbestos and lead paint are well documented. Other hazardous materials include arsenic-impregnated timber and formaldehyde binders in insulation foams, and pressed-wood products.

Closing the health equity gap

- **Climate-friendly housing policies and initiatives should focus greater attention on the 40% of urban growth in slums.** Planned design of residential communities should emphasize safer siting, energy-efficient construction, and development in proximity to basic services accessible by walking/cycling as well as public transport networks. Simple innovations such as insulated roofs, low-energy/solar lights, solar water heating, and improved biomass stoves can improve health equity and reduce health impacts from extreme weather.
- **Energy-efficient biomass and gas cookstoves in developing countries** can help avert more than 1 million deaths from chronic obstructive pulmonary disease (COPD), mostly among poor women exposed to indoor air pollution from stove emissions. Cleaner cookstoves also can help avert nearly 1 million deaths annually among children under the age of 5 from pneumonia caused by indoor smoke exposure.
- **Energy-efficient home heating also can generate health and equity benefits.** Passive solar "combi" home and hot water heating systems, biomass pellet stoves and heat pumps have been used in diverse low- and middle-income settings to raise indoor thermal comfort levels, and reduce respiratory illnesses. More energy-efficient heating can also help reduce risks of burns and injuries. Passive solar hot water heating can help improve kitchen hygiene.
- **Access to safe drinking-water and improved sanitation also** can be supported through greater emphasis on: domestic water conservation, grey water recycling, sanitation waste conversion to biogas, and low-carbon, on-site sanitation management where appropriate. Reducing water and energy costs associated with water extraction, consumption, contamination and poor wastewater management also can help improve access by ensuring the long-term resilience of water resources.
- **Air conditioning can reinforce health inequities** by exacerbating urban noise and heat, which negatively affect the health of others, particularly those who cannot afford air conditioners. Also, air conditioning contributes to climate change due to heavy energy consumption and use of powerful greenhouse gases as coolants.
- **Replacing kerosene lighting with solar-powered lanterns** can potentially reduce risk of injuries and eye diseases as well as indoor air pollution exposures in poor households of developing countries. Expanding access to DC (direct-current) household appliances (e.g. refrigeration, phones, computers) that can be powered directly by photovoltaic solar panels may improve health equity and help reduce climate change.
- **Many low-income cities are experimenting with healthy low-cost, climate friendly housing;** policies supporting such measures should be studied, expanded and evaluated.
- **Stronger building codes and housing finance measures can facilitate investment** in healthier and more energy-efficient housing, and help households avoid excessive fuel expenditures, or "energy poverty".
- **Improved international and national climate finance mechanisms** are needed to help fund measures that support healthy housing, particularly for the poor.

Background and rationale

This analysis reviews and evaluates the potential health impacts of mitigation strategies and technologies for the residential building sector, with a focus on strategies reviewed in: *Mitigation of Climate Change. Contribution of Working Group III to the Fourth Assessment Report of the Intergovernmental Panel on Climate Change,* also referred to here as the *IPCC mitigation review*.[ii]

Residential buildings are responsible for nearly 18% of direct carbon dioxide emissions (International Energy Agency, 2008), with 11% due to household grid consumption of electricity and district heating, and the remainder from household-level cooking and heating (e.g. with natural gas, LPG or biomass/coal). The residential and commercial building sector was described by the IPCC mitigation review as having the greatest potential for reducing greenhouse gas (GHG) emissions cost-effectively, within a short time using available and mature technologies. This is in comparison to other IPCC-assessed sectors including transport, agriculture, industry, forestry, energy supply and waste generation. Housing is therefore a significant factor in greenhouse gas emissions and climate change.

At the same time, housing and the built environment have a profound impact on human health. Healthy housing conditions can significantly decrease risk of communicable and noncommunicable diseases. Demographic and migration trends mean that the world's urban population will double by 2050, with most urban growth occurring in low- and middle-income cities. That, in turn, translates into an explosion of urban housing construction and/or informal settlement and slum expansion. Clearly, then, the way in which new housing is developed will have far-reaching impacts on urban health risks – as well as on urban safety, energy efficiency, heat wave resilience, access and mobility, and other urban health determinants.

Not all mitigation measures, however, have identical health impacts. Some measures may be highly positive for health, while others may generate new and unforeseen health risks if simple preventive measures are not incorporated. For instance, insulation improvements in temperate climates need to include measures to ensure adequate ventilation so as to avoid transmission of airborne infections, such as tuberculosis, or accumulation of indoor air pollutants, including toxic chemicals and radon. At the same time, low-energy buildings in warm climates and malaria-endemic regions that include design features to promote cooling with natural ventilation need to consider screening or other measures to protect against vector-borne diseases. Health-informed choices between mitigation measures in housing and construction can thus significantly impact strategies, but also ensure the best benefit-to-cost ratio for investments made by reducing concrete costs of disease and injury and also improving public health.

---

ii Levine M, Urge-Vorsatz D. Residential and commercial buildings In: Metz, B et al. eds. *Climate Change 2007: Mitigation of Climate Change. Contribution of Working Group III to the Fourth Assessment Report of the Intergovernmental Panel on Climate Change.* Cambridge University Press, Cambridge and New York, 2007.

Scope and methods

This report looks first at the climate and environmental impact of housing (Chapter 1) and then at how housing impacts health (Chapter 2) with respect to building siting and land use, choices of construction materials, design features, ventilation and energy, and also inhabitant behaviour. Summaries of key evidence are presented in two categories, which often overlap:

- **Housing-related risks to health, such as:** poor indoor air quality – e.g. indoor smoke from heating and cooking, moulds and moisture, exposure to carcinogenic or otherwise harmful chemical pollutants from building materials such as asbestos, lead and formaldehyde, as well as radon underground; thermal conditions – exposure to extremes of heat and cold; pests and infestations; noise; and urban design – which may facilitate or deter healthy physical exercise and childhood mobility.
- **Housing-related diseases and injuries, where significant evidence exists, including:** tuberculosis and other airborne infectious diseases, asthma, water-borne diseases impacted by lack of clean drinking-water and sanitation access, vector-borne diseases, home injuries and mental health.

Chapter 3 examines specific mitigation measures considered by IPCC alongside the body of evidence about the health impacts of housing. Identified health studies of specific intervention measures are given special attention, e.g. studies of health impacts of insulation and energy efficiency programmes. This brings the broad knowledge about housing and health into focus on measures proposed for climate mitigation.

While the IPCC assessment covers both residential and commercial buildings, this analysis was limited to residential buildings.

IPCC-reviewed measures that were considered included strategies for: improving the "thermal envelope"[iii] of buildings; use of more energy-efficient heating systems; use of passive solar systems for heating and domestic hot water production; reduction of building cooling requirements ("cooling load")[iv]; design and landscaping; and ventilation measures. Also considered are measures for heating, ventilation and air conditioning (HVAC) systems; daylighting and lighting, including photovoltaic solar panels for electricity generation; and certain efficiencies in household appliances.

Summary of findings

Climate change mitigation strategies in the housing sector can yield both immediate health gains and long-term mitigation objectives, as long as the choice of measures to be adopted explicitly considers potential health benefits and risks. Health inequalities can be addressed by deploying low-carbon climate change mitigation measures adapted to slums and other poor communities. Implementation of climate change mitigation

---

[iii] Thermal envelope refers to the shell of the building as a barrier to unwanted heat or mass transfer between the building interior and outside conditions. (Source: IPCC *Working Group III – Fourth Assessment Report*).

[iv] The hourly amount of heat that must be removed from a building to maintain indoor comfort, measured in British thermal units (BTUs). (Source: US EPA, *Terms of Environment: Glossary, Abbreviations and Acronyms*; http://www.epa.gov/OCEPATERMS/bterms.html)

measures should consider occupational health risks and relevant exposures of workers engaged in construction or retrofits of homes. Home occupant behaviour should also be considered, as it influences the effectiveness of certain mitigation measures and impacts on health (e.g. regulation of indoor temperature and ventilation). Findings reflect an urgent need for including health in housing policies, for example in improved building standards and enforcement of housing codes. Climate-related finance and other housing finance mechanisms should consider the health benefits and risks of climate-friendly construction or retrofits alongside carbon savings. More careful evaluation of potential health benefits and risks from all strategies, as well as monitoring and follow-up of their impacts, can ensure "win-win" outcomes for health and environment in accordance with the following principles:

- **Consider health co-benefits and risks at the planning stage.** Health impact assessment (HIA) of proposed housing climate change mitigation strategies can be applied to a specific intervention or package of measures. This can provide information about the expected health impacts of alternative scenarios as well as practical recommendations to improve the health performance of climate change mitigation strategies.
- **Ensure that housing strategies include land use and transport planning for walking, cycling and rapid transit/public transport, as well as access to green areas** to enhance health and climate benefits and reduce risks (e.g. urban heat island effect).
- **Ensure that appropriate standards and codes are in place, particularly to safeguard basic structural features** such as access to electricity, safe drinking-water, proper sanitation, natural ventilation and lighting, and to avoid use of materials with health hazards.
- **Develop/use healthy housing criteria, checklists and good practice guidance** to select strategies and investments and to monitor healthy housing indicators.
- **Document reductions in risks to health, benefits to health and savings in health care costs** related to housing interventions; this information is useful in communicating health gains and related savings.
- **Build capacity of health and non-health professionals** regarding mitigation measures and their potential health impacts using a systems approach that considers GHG impacts at all stages of building construction and use.

The main findings of this review are summarized in Table 1. The potential for mitigation strategies to provide health co-benefits or generate health risks are classified as: -- (strongly negative health impact); – (negative health impact); + (positive health impact); ++ (strongly positive health impact). These are weighted classifications relating to two factors: 1) qualitative evaluation of the evidence based upon expert opinion, as well as 2) number and quality of scientific studies available (e.g. study design, sample size, and consideration of potential confounding factors). These classifications should be regarded as indicative rather than definitive.

Table 1. Appraisal of health implications of selected mitigation strategies

Mitigation strategy	Likely health co-benefits		Impact of health co-benefit	Health risks to be avoided	Impact of health risk
Improved thermal performance of building envelope (IPCC 6.4.2)	**Environmental exposure**			Risk of inadequate ventilation: a) Reduced indoor air quality leading to potentially increased concentrations of indoor air pollutants (e.g. radon, mould and moisture) as a cause of asthma, bronchial obstruction and other illnesses b) Increased airborne infection transmissions (e.g. TB); risk of exposure to health-damaging insulation materials and fibres that cause cancer and other illnesses	- -
	Thermal comfort		++		
	Noise exposure reduction		+		
	Disease risk reduction				- -
	Reduced cardiovascular diseases, bronchial obstruction, asthma and other respiratory conditions		++		
	Reduced vector-borne disease due to infestations and pests		++		
	Better mental health through thermal comfort		+		
	Equity impacts				
	Depends on access of poor to improvements		+		
Low-carbon-emissions heating systems and passive solar design (IPCC 6.4.3, 6.4.6–7)	**Environmental exposure**			Field studies have found that more cost- and energy-efficient heating do not always reduce net household energy use (and thus energy-related greenhouse gases and air pollutants) by an equivalent amount. This is because some households may allocate a portion of their cost savings to *increase* their energy (electricity or heat) consumption, a phenomenon described as the "take-back effect"	0
	Thermal comfort		++		
	Hygiene		+		
	Disease risk reduction				
	Reduced asthma and respiratory symptoms related to cold exposure, damp and mould		++		
	Reduced pneumonia and COPD (in case of reduced biomass use)		++		
	Better mental health due to better thermal comfort		+		
	Equity impacts				
	Depends on access of poor to improvements		+		
Reduced cooling loads on buildings through design features and improved natural ventilation (IPCC 6.4.4)	**Environmental exposure**			May not work when night temperatures remain high; need to be adapted to regional humidity	0
	Thermal comfort		++		
	Disease risk reduction			Design must take account of winter as well as summer risks	0
	Reduced asthma/respiratory illness from particulates, radon, mould, etc.		++		
	Reduced TB and other airborne infection transmission risk		++	Natural ventilation without house screening may increase vulnerability to vector-borne diseases	- -
	Less airborne disease transmission via air-conditioning systems		+	May increase exposure to high outdoor air pollution concentrations, causing respiratory symptoms, unless filters are used	- -
	Equity impacts				
	High equity co-benefit from broader access to effective cooling and ventilation, particularly when design measures are adopted in low-income settings		+	Avoid use of lead in paint (e.g. white paint for albedo effect)	-

Strongly positive health impact ++; Positive health impact +; Strongly negative health impact: - - ; Negative health impact: -

Mitigation strategy	Likely health co-benefits		Impact of health co-benefit	Health risks to be avoided	Impact of health risk
More energy-efficient and better-maintained heating, ventilation and air conditioning systems (HVAC) Greater reliance on building design and natural ventilation (IPCC 6.4.4–5)	**Environmental exposure**	Thermal comfort	++	Greater risk of airborne infectious diseases (e.g. tuberculosis) and upper and lower respiratory symptoms in AC rooms/spaces lacking sufficient fresh air exchanges	- -
		Reduced noise exposure	+		
	Disease risk reduction	In settings with significant outdoor air pollution, reduced respiratory symptoms and asthma	++	Increased urban dependence on AC stimulates vicious cycle of exacerbated urban heat island effect	-
		Less risk of cardiovascular disease due to heat exposure	++	More noise and pollution exposure for those not using air conditioning	-
		Less risk of vector-borne disease due to closed windows	+	Bacterial proliferation/legionellosis in very large HVAC tanks/cooling towers	- -
	Equity impacts	Those least able to afford AC suffer the most from its noise and heat island impacts.	-	Delayed climate-related health impacts from added greenhouse gas emissions of air conditioners	-
Passive solar hot water and photovoltaic solar electricity (IPCC 6.4.7–8)	**Environmental exposure**	Hygiene and sanitation	+	Greater initial cost outlays pose barriers for poor families if not offset by subsidies	-
	Disease risk reduction	Less asthma and respiratory disease due to decreased use of kerosene lighting in developing countries	+	New technology risks require more assessment, including of occupational and environmental risks of production and exposure to waste byproducts, e.g. respiratory irritations and impacts of exposures to heavy metals or other toxic substances.	0
		Fewer burns from kerosene appliances	+		
	Equity impacts	More access to electricity among poor and rural populations	++		
		Lower long-term electricity cost once initial investment is made	+		
Lighting and day lighting: window positioning to reduce heat/cold impacts; highly energy-efficient indoor lighting (IPCC 6.4.9–10)	**Environmental exposure**	Thermal comfort	++	Household injury from inadequate indoor/proximity lighting	-
	Disease risk reduction	Less asthma and respiratory disease due to natural ventilation through windows	+		
		Fewer home injuries (falls)	++		
		Positive effect of light on metabolic function and mental health	+		
Household appliances and electronics: more low-energy and direct-current appliances, including improved biomass cookstoves (IPCC 6.4.11; 6.6.2)	**Environmental exposure**	Reduced indoor air pollution	++	Equity gains dependent on increased access of poor to new low-energy cookstove technologies and other appliances	-
		Improved food safety, kitchen hygiene	+		
	Disease risk reduction	Reduced asthma and respiratory disease	+	In developed countries, more efficient appliances may not decrease GHG and air pollution emissions if there is not a equivalent decrease in overall energy use	-
		Fewer injuries from burns due to inadequate cooking and heating appliances	++		
		Less COPD, cancer and cardiovascular disease	+		
	Equity impacts	Access to cleaner biomass and biogas cookstoves	++		

He Jianqing

Pinggu District, China: Photovoltaic (PV) solar-powered lights illuminate the streets of this new neighborhood in the Beijing region, and solar-embedded rooftop panels support

Introduction

Background and rationale

Residential buildings contributed close to 18% of direct carbon dioxide emissions from energy combustion in 2008, with 11% due to household use of grid electricity and district heating, and the remainder due to emissions at household level (e.g. cooking and heating with gas, coal, oil, etc.).

The residential and commercial building sector has been described by the *Intergovernmental Panel on Climate Change*[i] as having the greatest potential for reducing greenhouse gas (GHG) emissions cost-effectively within a short time, using available and mature technologies. This is in comparison to other IPCC-assessed sectors including transport, agriculture, industry, forestry, energy supply and waste generation. At the same time, IPCC notes that in a high-growth scenario, total building-related greenhouse gas emissions could nearly double by 2030, with developing country emissions exceeding those in North America, Europe and the Caucasus and Central Asia regions. Housing is therefore a significant factor in greenhouse gas emissions and climate change.

Housing and the built environment in general have profound impacts on human health. Healthy housing can significantly decrease communicable and noncommunicable disease risks. At the same time, public health is impacted by the vulnerability of housing environments to climate change effects such as flooding and extreme weather. This leads to a vicious cycle that can only be broken by integrated mitigation efforts that address housing, environment and health linkages.

Additionally, the world's urban population will nearly double by 2050, increasing from about 3.3 billion people in 2007 to about 6.4 billion, and most of that growth will take place in low- and middle-income countries.[ii] Therefore, housing is an important entry point for addressing critical urban health concerns. Also, the home's role in health is all the more important to vulnerable population groups (the poor, sick, children, elderly and disabled) who spend comparatively more time in this setting and are particularly in need of healthy living environments.

---

[i] Levine M, Urge-Vorsatz D. Residential and commercial buildings. In: Metz B et al., eds. *Climate Change 2007: Mitigation of Climate Change. Contribution of Working Group III to the Fourth Assessment Report of the Intergovernmental Panel on Climate Change.* Cambridge University Press, Cambridge and New York, 2007.

[ii] *World Urbanization Prospects, the 2007 revision. Executive Summary.* New York, United Nations, 26 February 2008. accessed at: http://www.un.org/esa/population/publications/wup2007/2007WUP_ExecSum_web.pdf .

Scope and methods

This analysis reviews potential health impacts of mitigation strategies and technologies for the residential building sector, focusing on strategies considered in IPCC's *Mitigation of Climate Change: Contribution of Working Group III to the Fourth Assessment Report*.

While most focus is placed on mitigation strategies considered by the IPCC, some strategies not mentioned by the IPCC are considered. These strategies take advantage of opportunities offered by urban development, behavioural change and other factors to generate health and environment co-benefits.

> This analysis reviews the health impacts of mitigation in housing considering the physical house design and structure, as well as the neighborhood and the community.

Mitigation options of household energy systems and health co-benefits are considered separately in another report in this series (*Household energy in developing countries*) with particular reference to emissions-intensive biomass and coal-burning systems in developing countries. Another report in this series (*Health care facilities*) deals with IPCC-reviewed mitigation options for buildings that are a primary site for health sector activities.

Parameters of health and housing considered

WHO has adopted a broad definition of healthy housing that refers to four related dimensions: the house structure, the home social environment, the neighbourhood and the community. In light of existing knowledge, healthy housing is regarded as a means of protecting inhabitants' health from a variety of risks in the built and natural environment – physical, chemical, biological and psycho-social.

WHO thus undertook this review of potential health co-benefits and risks of mitigation strategies relevant to buildings with reference to the mitigation strategies considered by the IPCC. While the IPCC review covered both residential and commercial buildings, this review is limited to residential buildings.

Scope of literature review

Mitigation strategies were reviewed in light of evidence relating to three key parameters for potential health impacts (co-benefits or risks):

- Impacts on housing-related health risks to inhabitants, construction workers and/or the community;
- Impacts on specific communicable and noncommunicable diseases, including home injuries and mental health diseases/conditions, for inhabitants/construction workers and/or the community;
- Impacts on health equity and access to healthy housing conditions.

Overall, IPCC refers to three main principles for reducing building-related emissions: increasing buildings' energy efficiency, reducing energy use and shifting to renewable energy sources. In this context, the IPCC-reviewed mitigation strategies appraised in terms of their health co-benefits and risks included the following:

1. Improvement of the thermal envelope of buildings (IPCC 6.4.2)
2. Heating systems, including passive solar thermal measures (IPCC 6.4.3; 6.4.6–7)

3. Cooling loads (IPCC 6.4.4)
4. Air conditioning and heating, ventilation and air conditioning systems (HVAC); (IPCC 6.4.4–5)
5. Passive solar water heating and photovoltaic solar electricity (IPCC 6.4.7–8)
6. Lighting and day lighting (IPCC 6.4.9–10)
7. Household appliances (including cookstoves) and electronics (IPCC 6.6.2; 6.4.11)

Search strategies of databases and literature review

Overall literature review

Current knowledge about key health risks associated with housing conditions was reviewed. Also documented was the climate change mitigation potential of key mitigation strategies relevant to residential buildings. This focused on peer-reviewed literature in engineering, architecture and design, as well as assessment of energy efficiencies gleaned from development, water and sanitation, engineering and architectural reviews. In addition to the resources included in the above mitigation matrix, approximately 100 additional sources were used throughout chapters 1, 2, 5, 6 & 7.

Core analysis of health co-benefits from mitigation strategies

The potential health impacts of the seven key IPCC-reviewed strategies and additional strategies were appraised in the light of health-based evidence of co-benefits and risks as identified in a review of nearly 120 peer-reviewed published articles and reports, including reports by bilateral and multilateral development agencies published in English between 1980 and 2010. In addition to health evidence, additional research on mitigation strategies (beyond that reviewed by IPCC) was considered where relevant. References were also gained from a 13–15 October 2010 meeting in Geneva of 40 international housing, health and climate change experts organized by WHO.

Given the breadth of the topic, risk factors and the health outcomes, we undertook a scoping review to summarize key findings in existing literature and identify major gaps where applicable, rather than conducting systematic reviews on each of the IPCC housing climate change mitigation categories. The results of the review should thus be regarded as indicative rather than definitive.

Based on the literature review, the strengths of likely health effects of a given mitigation strategy or package of strategies are described and classified from : -- (strongly negative health impact); - (negative health impact); + (positive health impact); ++ (strongly positive health impact). These are weighted classifications relating to two factors: 1) qualitative evaluation of the evidence based upon expert opinion, as well as 2) number and quality of scientific studies available (e.g. study design, sample size, and consideration of potential confounding factors, etc.). These classifications, presented in Table 7 (Chapter 4), should be regarded as indicative rather than definitive.

Inclusion/exclusion criteria

Peer-reviewed literature (see summary table, p. 13) included epidemiological studies and other intervention, observational and case studies which examined impacts or associations between health and housing design, building features, energy use, building materials, etc. The analysed studies focused on evidence about housing-related non-communicable and communicable diseases, where available, and housing and health risks, e.g. indoor air pollution, home injury, noise, etc. (see Chapter 2). Systematic reviews of literature were identified and highlighted wherever possible. International (e.g. United Nations system; International Energy Agency) and multilateral or national reports of trends and statistics were referenced as appropriate. Civil society reports were used to illustrate experiences and approaches to some of the problems identified by the review. Studies funded by commercial interests, not subject to independent peer review, were excluded.

Limitations

In many cases, "matching" relevant health and mitigation evidence posed challenges. Mitigation evidence addresses building or design strategies for climate change in ways that differ from health evidence categories. For instance, while the mitigation literature may deal with the "thermal envelope" effect of improving insulation and energy efficiency, health literature relates to actions taken to "improve thermal conditions." Often this may include a mix of interventions, and without explicit reference to whether the measure was more or less energy-efficient or how much. However, examples like the studies undertaken in New Zealand reflect pioneering work on health co-benefits along with reductions in energy consumption and emissions through insulation and energy efficiency interventions.[iii, iv]

[iii] Chapman R et al. Retrofitting houses with insulation: a cost-benefit analysis of a randomised community trial. *Journal of Epidemiology and Community Health*, 2009, 63(4):271–277.

[iv] Howden-Chapman P et al. Effect of insulating existing houses on health inequality: cluster randomised study in the community. *British Medical Journal*, 2007, 334:460.

IPCC-reviewed mitigation strategy	Databases searched	No. health studies reviewed	Sources of studies included	Types of studies included
1. Improvement of the thermal envelope of buildings	PubMed, Google Scholar	28	International organization reports, national government agency reports and peer-reviewed literature	Cost-benefit analysis Intervention studies Epidemiological assessment Systematic reviews Guidelines Randomized control trials
2. Heating systems, passive solar systems and domestic hot water	PubMed, Google Scholar	16	National agency reports and peer-reviewed literature	Randomized control trials Cost-benefit analysis Intervention studies Epidemiological assessment Systematic reviews
3. Cooling loads	PubMed, Google Scholar	17	International organization reports and peer-reviewed literature	Decision-making analysis Epidemiological assessment Guidelines Intervention studies
4. Air conditioning and heating, ventilation and air conditioning systems	PubMed, Google Scholar	21	International organization reports and peer-reviewed literature	Decision-making analysis Epidemiological assessment Intervention studies
5. Photovoltaic solar energy for electricity generation	PubMed, Google Scholar	6	Peer-reviewed literature	Case studies Systematic review
6. Lighting and day lighting	PubMed, Google Scholar	6	National agencies and peer-reviewed literature	Guidelines Intervention studies
7. Household appliances and consumer electronics	PubMed, Google Scholar	7	Peer-reviewed literature	Intervention studies Epidemiological assessment

Additional strategies identified	Databases searched	No. of studies included	Sources of studies included	Types of studies included
Healthy urban design	PubMed, Google Scholar	16	International organization reports and peer-reviewed literature	Case studies Policy briefing Epidemiological assessment
Behavioural changes	PubMed, Google Scholar	4	International organization reports and peer-reviewed literature	Case studies Systematic review Epidemiological assessment

1

FAO / Giuseppe Bizzarri

Caracas, Venezuela: Urban gardens like this one provide multiple health benefits: local fresh produce, physical activity and green spaces around housing to offset the urban "heat island" effect.

Overview of housing and climate/environment linkages

This chapter presents a brief overview of critical aspects of housing's impact on climate change trends and, conversely, how climate change affects housing environments in ways that are relevant to health.

Issues addressed here include global trends in housing emissions, trends in developed versus developing and emerging economies, and issues related to urban form such as housing density and slum growth. Also addressed briefly are climate change impacts on housing and vulnerabilities in the context of housing environments, and how these may vary in hot/cold and wet/dry geographic regions. These issues provide the context for discussion of the health impacts of housing developed in Chapter 2.

These trends share one overarching reality: 60% of the global population will live in cities by 2030[1], with most of population growth occurring in developing cities. This is where most new housing will likely be built. And the overall balance of health risks and benefits associated with housing environments will play a determining role in the health of the world's urban residents. Housing built over the next two decades will also play a critical role in climate change trends that impact on health.

1.1 How housing contributes to climate change

According to International Energy Agency data (2008), global residential emissions of carbon dioxide (CO_2) account for about 17.8% of total global direct CO_2 emissions from combustion sources. Of that total, 11.3% is housing-related grid electricity and district heating use, while 6.5% of CO_2 emissions are generated at household level, e.g. use of LPG/gas, coal and oil for cooking and heating[2] (Fig. 1).

This estimate relates to CO_2 emissions from combustion only. It thus does not consider climate change pollutants that may have an even

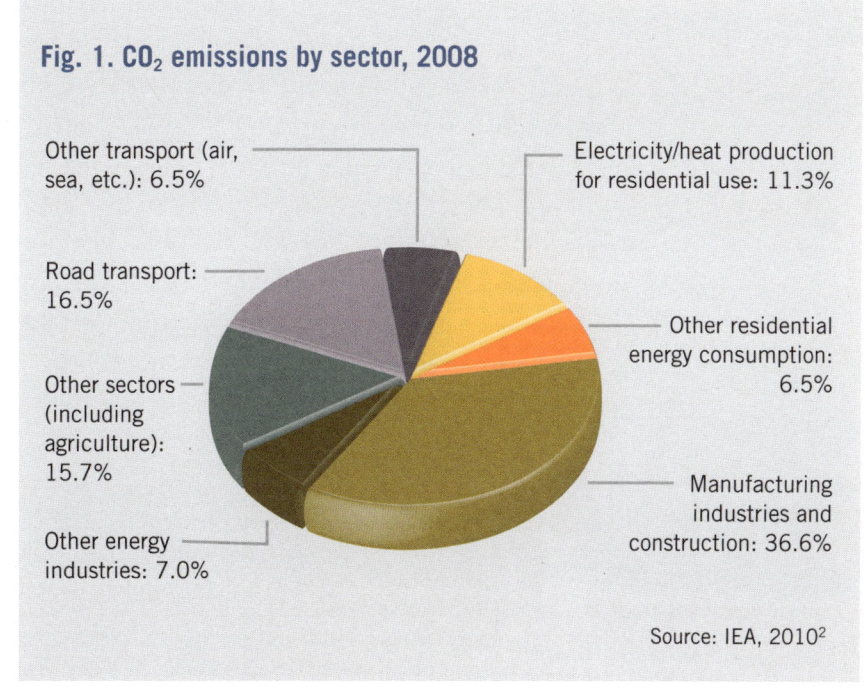

Fig. 1. CO_2 emissions by sector, 2008

Other transport (air, sea, etc.): 6.5%
Road transport: 16.5%
Other sectors (including agriculture): 15.7%
Other energy industries: 7.0%
Manufacturing industries and construction: 36.6%
Other residential energy consumption: 6.5%
Electricity/heat production for residential use: 11.3%

Source: IEA, 2010[2]

more powerful global warming potential than CO_2, such as methane and black carbon emitted by household biomass combustion in developing countries. Also excluded are refrigerants, powerful climate change agents used for home appliances and air conditioners. While a more complete accounting of emissions is thus needed, the data reflect the significant contribution of the housing sector to climate change.[i]

In the countries of the Organisation of Economic Co-operation and Development (OECD), residential and commercial buildings may be responsible for as much as 30% of primary energy consumed as well as 30% of OECD greenhouse gas emissions. Building-related energy use in OECD countries has continually increased since the 1960s.[3]

The *Fourth Assessment Report of the Intergovernmental Panel on Climate Change* projects that in a high economic growth scenario, building-related greenhouse gas emissions could nearly double by 2030 in the absence of firm mitigation measures.[4]

On the more positive side, the residential and commercial building sector is described as having "the highest immediate mitigation potential in terms of absolute reductions in CO_2-eq emissions that could be attained by the year 2030 at a cost of less than US$ 100 per ton of CO_2-eq." This is in comparison to reductions that, according to the IPCC review, could reasonably be achieved in sectors such as transport, agriculture, industry, forestry, overall energy supply and waste management.

1.2 Trends in developed versus developing countries

Historically, most building-related emissions were generated in North America, Europe, and certain regions of the Caucasus and Central Asia. Building-related emissions are increasing in many developing countries, and would account for most emissions growth in a high economic-growth scenario, according to IPCC. In slower-growth scenarios, emissions increases would be largest in Asia and North America.[4]

Per capita building energy consumption varies widely in high-, medium- and low-income countries (Fig. 2). But as review of health co-benefits will underline, it is important to consider how building energy use is influenced by a much wider range of factors beyond economic development levels, including: urban form and building characteristics (old, new, quality of thermal shell), regional climate variation, energy and building policies and, ultimately, human behaviour.

The composition of emissions is also important. For instance, biomass and coal combustion generate significant shorter-lived climate-change pollutants, including black carbon particles that are harmful to health.[5,6] Some experts believe that addressing these shorter-lived pollutants immediately can help slow the pace of climate change.[7,8]

---

[i] Chapter 1 of the *Contribution of Working Group III to the Fourth Assessment Report of the Intergovernmental Panel on Climate Change* (Rogner, H et al, 2007) provides another estimate of total greenhouse gas emissions by sector (Fig. 1-3b) in which 7.9% of CO_2-eq. emissions are attributed to residential and commercial buildings (2004). This estimate, however, does not include emissions from building consumption of grid electricity and heating (attributed instead to the energy sector). It also does not consider certain climate change pollutants increasingly recognized as important, including black carbon from biomass combustion.

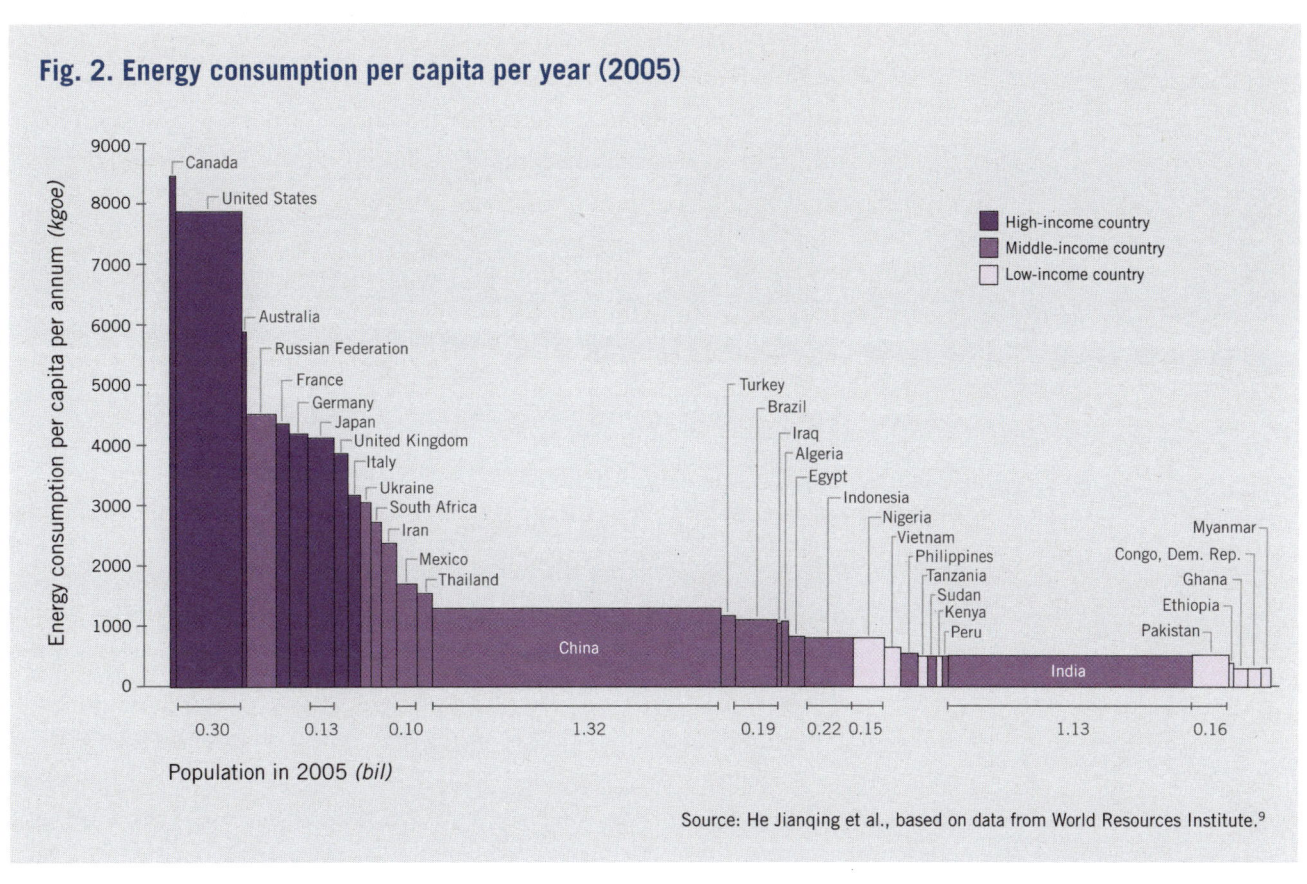

Fig. 2. Energy consumption per capita per year (2005)

Source: He Jianqing et al., based on data from World Resources Institute.[9]

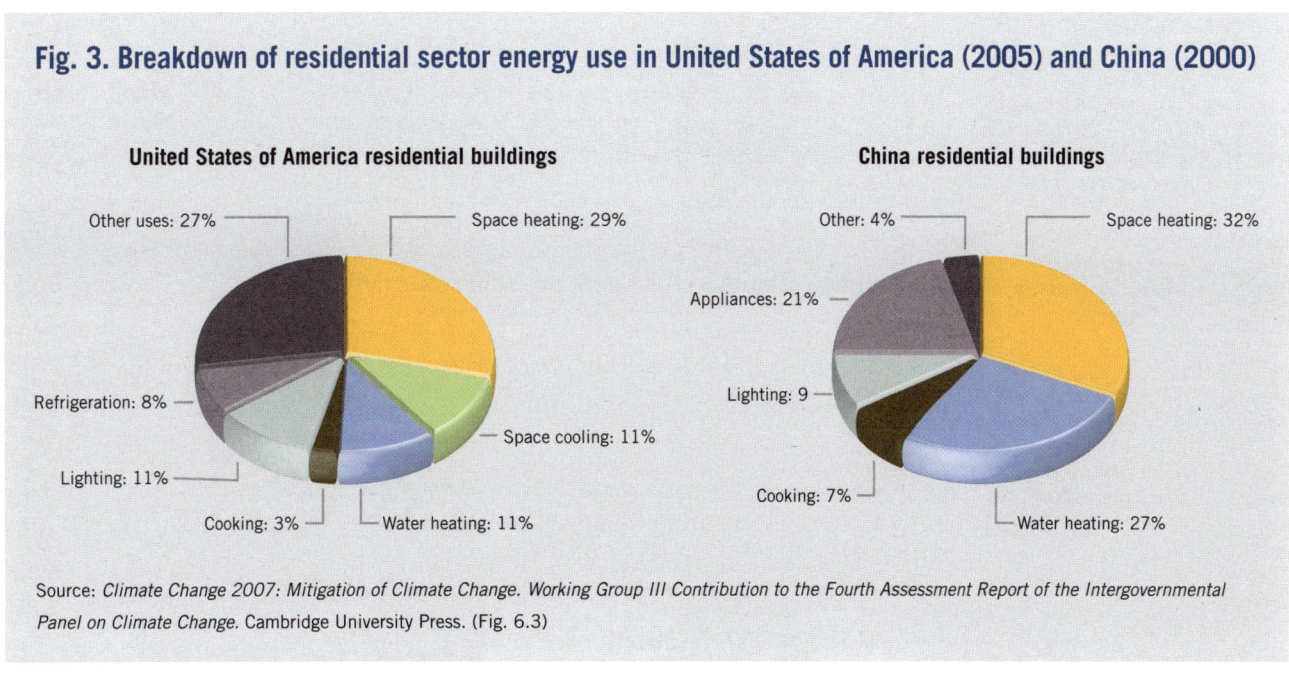

Fig. 3. Breakdown of residential sector energy use in United States of America (2005) and China (2000)

Source: *Climate Change 2007: Mitigation of Climate Change. Working Group III Contribution to the Fourth Assessment Report of the Intergovernmental Panel on Climate Change.* Cambridge University Press. (Fig. 6.3)

Typically, developing countries have lower energy consumption per unit of space than developed countries, but a larger proportion of their energy use is directly attributable to cooking, heating and lighting compared to appliances, with energy for heating being a climate-dependent variable.

The allocation of energy for different residential building uses is illustrated by IPCC for the United States of America and China (Fig. 3).[10] In both countries, the largest single use of energy is space heating, as is typical of most temperate regions. In China, however, water heating ranks second in terms of energy consumption, while in the USA, small electric appliances ("other") are the second most important energy consumers. This reflects the growth potential of electrical appliance consumption in emerging economies as more households gain access to electricity. Lighting and cooling are similarly important as the third- and fourth-largest energy consumers in both China and the USA countries.

On the other end of the scale, some 1.4 billion people in developing countries, or over 20% of the world's population, still lack access to any electricity at all. A WHO/UNDP joint report on energy access found almost 100% of people in OECD and transitional economies have access to electricity, while only 72% of people in developing countries have such access (Table 2).[11] Almost half of developing countries (68 of 140) have established targets for access to electricity.

Table 2. Access to electricity in the world, 2008

	Total population (in millions)	Electrification rates (%)	Total population without electricity (in millions)
World	6692	78.2	1456
OECD and transitional economies	1507	99.8	3
Developing countries	5185	72	1453

Source: UNDP/WHO, 2009[11]

Some 3 billion people, mostly in developing countries, also rely upon biomass or coal fuels for most of their home cooking and heating needs. This represents about 46% of the global population[12]. Very few countries have set targets for access to cleaner forms of household fuels (17 countries), and fewer still have set targets for access to improved household biomass cooking stoves (11 countries). This issue is discussed in a companion report in this series: *Health in the Green Economy: The Household Energy Sector in Developing Countries.*[13]

Until recently, home air conditioning, while common in warm regions of many developed countries, was largely limited to large office buildings, hotels and high-income homes of the developing world. The IPCC review, however, notes: "That is quickly changing, however, with individual apartment and home air conditioning becoming more common in developing countries, [and] reaching even greater levels in developed countries. This is evident in the production trends of typical room- to house-sized units, which increased 26% (35.8 to 45.4 million units) from 1998 to 2001."[10]

1.3 Housing density and urban design as factors in GHG emissions

In developed countries, mid- to high-density housing tends to be more energy-efficient than low-density housing of comparable size and standards. Residents of denser settlements are thus likely to have lower overall per-capita emissions than residents of surrounding areas as a result of building and urban efficiencies, e.g. greater use of public transportation systems.[14, ii]

A study comparing a high-density multi-storey condominium project near Toronto's urban core and a low-density residential neighbourhood on the city's suburban fringe illustates how urban residential density may impact energy efficiency and greenhouse gas emissions. The study took a "life cycle" approach, assessing emissions associated with infrastructure development, building construction and use/operation, as well as transportation patterns. Low-density development was found to generate roughly 2.5 times the annual GHG emissions on a per-capita basis as the high-density development. Similarly, the low-density suburban neighbourhood used approximately twice as much energy annually, per capita, as the higher-density development.[15]

Housing densities, which vary by country, culture and region, as well as economies, can have a range of impacts on both health and environment. Pictured here is an apartment building in Nagoya, Japan. (Photo: Andrew Martin)

At the same time, very high-rise apartment buildings may generate more greenhouse gas emissions if they rely upon very large heating and cooling systems, and/or other electronic/mechanized features. Energy-efficient single-family units also may produce emissions well below average (although energy for transport remains a factor). Thus emissions vary greatly by dwelling type, geographic location and other factors.

In low-income countries, where urban populations are increasing the most rapidly, the urban pattern of energy use may be somewhat different. Greenhouse gas emissions may still be comparatively higher in urban centres – where per-capita income and access to electricity are often higher, as compared to peri-urban or rural areas.

At the same time, low-income cities are growing "horizontally" into megacities or mega regions.[16-17] The periphery of many cities may include low-density suburbs for more affluent groups, which imitate developed-country styles and introduce energy-intensive patterns of transport and infrastructure delivery. Low-rise, informal settlements of poor, migrants and (in conflict or post-conflict areas) refugee camps comprise another form of urban sprawl.[iii] Such settlements often have little access to energy, safe drinking-water and sanitation. And as they become more established, needs grow ad hoc, often without adequate infrastructure provision. For urban leaders, addressing urban health and environmental challenges in such a context is enormously challenging.

---

ii Glaeser and Kahn[14] studied new construction across the United States of America and found strong associations between zoning, land use and carbon emissions. Cities with higher densities and mixed residential/commercial use zoning generally had significantly lower emissions than suburban areas, characterized by lower densities and strictly separated residential and commercial zones. The city/suburb gap was particularly pronounced when the urban areas concerned were older and featured a very strong mixed-use zoning and transit orientation, such as New York City.

iii In conflict and post-conflict zones, refugee camps created ad hoc and during wars or major natural disasters may over decades develop permanent urban features, creating a range of housing and health challenges for countries in sub-Saharan Africa, Asia and the Eastern Mediterranean region, as well as in post-conflict zones of Europe.

Additionally, if horizontal expansion of cities is in "single function" residential or commercial zones, these areas become difficult to serve with public transport and require more roads and parking spaces for private vehicle transport.

The result may be peri-urban areas swathed by large areas of asphalt, which amplify the urban "heat island" effect (See 1.5.2). As efficient and safe travel via public transport, walking and cycling becomes more difficult, use of, and dependence on, travel by private vehicles is reinforced, adding to pollution, noise and injury risks.

Many low-income cities also have seen a surge in travel by motorcycles and three-wheelers, making safe, healthy movement by walking and cycling even more dangerous for the user, as well as inefficient for the entire transit system. These issues, and their relationship to health, are described more in the companion report from this series: *Health in the Green Economy: health co-benefits of climate change mitgiation in the transport sector*.[iv] The point here is that urban development styles and densities have a profound impact on health in the broader housing environment.

1.4 Slums and their environmental/climate change impacts

About 38% of the world's urban growth is occurring in slums (Fig. 4). A new report published in November 2010 by WHO and the United Nations Human Settlements Programme (UN-HABITAT) notes that nearly one billion people – one third of the urban population – are living in urban slums and shantytowns, stating: "slums are no longer just marginalized neighbourhoods housing a relatively small proportion of the urban population. In many cities, they are the dominant type of human settlement [...], carving their way into the fabric of modern-day cities, and making their mark as a distinct category of human settlement that now characterizes so many cities in the developing world."[17]

In Mumbai, India, new high-rise blocks contrast with low-rise slum dwellings and sprawl by the water. (Photo: © Acharya, Hard Rain Picture Library)

Slums are commonly defined as residential areas that lack one or more of the following: improved sanitation, safe drinking-water, security of tenure, durable housing and sufficient living area. Sufficient living area, as defined by UN-HABITAT, is no more than three people sharing the same room.[18] Slum conditions exacerbate many illnesses and disease conditions that are environmental health risks of housing in general. These risks include: high prevalence of diarrhoea from unsafe drinking-water and/or unimproved sanitation; high TB prevalence due to crowding and lack of adequate ventilation; exposure to vector-borne diseases such as malaria and dengue partly attributable to unsafe water, poor sanitation and waste management; risks of respiratory illnesses from indoor air pollution and mouldy housing interiors; and risk of traffic injury from

---

[iv] Hosking J et al. *Health in the Green Economy: health co-benefits of climate change mitigation in the transport sector*. Geneva, World Health Organization, 2011.

lack of access to safe public transport or safe walking and cycling routes to jobs and schools elsewhere. These risks are discussed further in Chapter 2.

1.5 Regional climate-related impacts on housing environments

Just as housing is a factor in greenhouse gas emissions, housing environments and health are impacted by climate change. This leads to a vicious cycle that can only be broken by more effective mitigation efforts addressing housing, environment and health linkages. While this section focuses on describing a few key environmental risks, related health conditions are described in more detail below.

Many housing- and climate change-related vulnerabilities require immediate attention in the context of adaptation measures. However, these measures can also be discussed in the context of mitigation efforts to make housing more energy-efficient and thermally protective. An integrated approach can reduce future health risks caused by climate change, and enhance health co-benefits enjoyed now and in the future.

1.5.1 Vulnerability to extreme weather events

Extreme weather events such as floods, heat waves and cold spells cause serious health and social problems all over the world. One of the clearest impacts of climate change on housing is apparent in the vulnerability of housing in coastal zones to typhoons, hurricanes, tsunamis and flooding in river plains. As extreme climatic events are likely to become more frequent, the siting, structural integrity and resilience of housing become more important in protecting health from climate change.

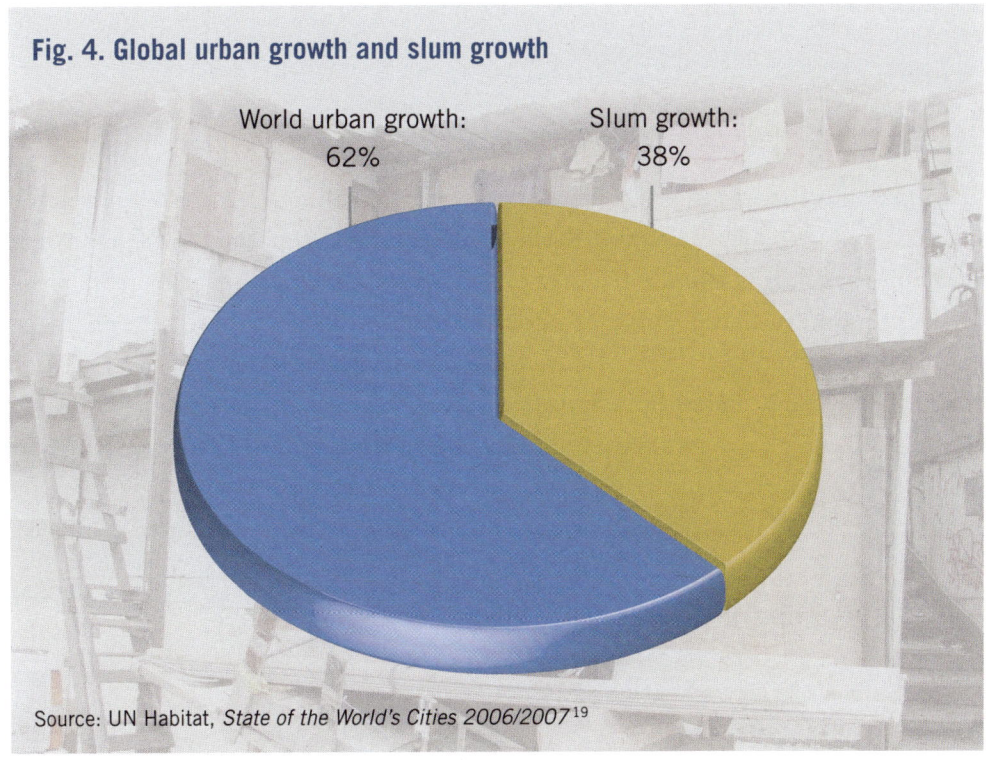

Fig. 4. Global urban growth and slum growth

World urban growth: 62%

Slum growth: 38%

Source: UN Habitat, *State of the World's Cities 2006/2007*[19]

Sometimes vulnerability may be exacerbated by local construction styles. For instance, mud brick houses may be more insulating against extreme heat than modern construction, but they are more vulnerable to structural collapse in heavy rains and earthquakes than structures built with modern techniques and stronger materials. Conversely, sometimes indigenous building styles (e.g. building on stilts to protect against flooding, or solid stone construction against extreme heat) may have been weakened or replaced by modern designs and need re-examination or adaptation.

Older buildings also may be less structurally resilient. For instance, one study reported that buildings erected since 1964 (about 32% of United Kingdom housing stock; see Table 3) are more resistant to climate-related extremes of heat and cold. Addressing such vulnerability issues in the context of mitigation strategies offers both opportunities to obtain co-benefits and challenges in situations where there may be trade-offs.

Table 3. A proposed vulnerability index for United Kingdom housing

Housing by age & building type		Pre-1919 26.4%	1919–45 19.7%	1945–64 21.5%	Post-1964 32.5%
Building type		**Vulnerability Index**			
Terraced houses	28.1%	1.32	1.15	0.79	0.40
Semi-detached	26.7 %	1.54	1.32	0.93	0.49
Bungalows	23.5%	2.0	1.72	1.21	0.65
Converted flats	6.9%	1.01	1.13	0.83	0.47
Low-rise flats	12.7%	0.81	0.7	0.48	0.25
High-rise flats	2%	0.49	1.42	0.29	0.17

Index: most vulnerable to climate impacts (2); least-vulnerable (0)
Source: United Kingdom Department of Health, 2001[20]

1.5.2 Urban 'heat island' effect

The urban 'heat island' effect refers to the disparity between urban and peri-urban temperatures, particularly acute during heat waves. This is a consequence of urban densification, sprawl, poor siting and design of built spaces, and the disappearance of green spaces in and around cities. It is exacerbated by climate change.[21, 22] The large unbroken expanses of built spaces and asphalt pavement characteristic of many urban areas absorb solar energy (visible and infrared radiation [IR]) and release huge amounts of heat as IR radiation. This can raise temperatures by 5–12°C compared to nearby rural areas.[23] The heat island effect is exacerbated in the absence of green spaces that otherwise filter heat-trapping air pollutants such as ozone. The heat island effect is a major factor in heat stress and excess mortality from heat waves, and thus reinforces reliance on air conditioning in a vicious cycle of increased housing-related GHG emissions.

1.5.3 Climate change impacts on indoor housing environments and indoor air quality

Extreme weather aside, a range of climate change-related factors affect indoor housing environments, and ultimately health, in powerful ways. Such factors include patterns of direct and diffuse sunlight, air temperature and movement, and amount and frequency of rainfall and related humidity. If climate factors change dramatically during a building's lifetime, it cannot provide appropriate indoor climate and shelter. These factors are considered in vulnerability assessment of housing and health in relation to climate change.

Here too, housing characteristics may provide greater or lesser degrees of resilience to such changes. Some critical factors include siting and density features (which, if properly planned, can provide added shade from heat), age of structures and types of building materials and energy systems that determine indoor air quality and respiratory health, as well as vulnerability to moulds and bacteria in humid climates.

1.5.4 Climate change vulnerabilities vary by eco-climatic region

Eco-climatic regions differ greatly in vulnerability. In arid desert regions, heat stress and indoor and outdoor air pollution from dust storms may be major housing-related issues. Yet cooler night temperatures in deserts offer the possibility of natural ventilation for cooling. Indoor dampness and moulds are widespread problems in many mid-continental, coastal and tropical regions; heavier seasonal rainfalls increase these problems.

Mould and dampness are estimated to affect 10–50% of indoor environments in Europe, North America, Australia, India and Japan.[24] Although little data are available for low-income countries, several studies suggest that indoor dampness is also common in settings such as river valleys and coastal areas. While mould can grow on all organic materials, selection of appropriate materials can prevent dirt accumulation and moisture penetration to retard mould growth. The urban heat island impact, while experienced in Europe, may be even greater in fast-growing developing cities near the equator, due to their large expanses of unshaded and paved areas. Yet the ample solar energy available to many developing cities offers relatively greater potential for solar energy use in housing, which along with passive design[v] can reduce energy-related housing emissions and pollution.

Domestic washing tasks are a challenge for residents of many slums and informal settlements where there is insufficient household access to water. Pictured here is a family in the Borde de Vias Colony in Mexico City, populated by recent migrants to the city. (Photo: © Mark Edwards, Hard Rain Picture Library)

1.5.5 Household water and sanitation

Safe drinking-water and sanitation are integral to healthy housing environments. Climate change impacts

---

v   Passive design is based on ensuring that the fabric of the building and the spaces within it respond effectively to local climate and site conditions in order to maximise comfort for the occupants.

can harm these systems. In coastal areas subject to frequent flooding, sewage systems, pipes and reservoirs may be vulnerable to inundation beyond their design capacities. Flooding increases the risk of bacterial dispersion and contamination of water resources, and may in turn lead to outbreaks of waterborne diseases. Significant changes in climatic conditions, such as higher temperatures and substantially changed rainfall patterns, can increase vector-borne diseases.

Climate impacts can compromise drinking-water safety as higher ambient temperatures increase bacterial growth and multiplication during distribution and storage. Similarly, treatment of household sanitation-related pollution is a major environmental issue in most cities, towns and villages in developed as well as developing countries.[25]

> Safe drinking-water and sanitation are integral to healthy housing. Well-designed water and sanitation services generate health and climate benefits, making provision of water infrastructure more energy- and cost-efficient, and therefore accessible.

Water and sanitation offer another example of the vicious cycle of local and global environmental change, particularly in water-scarce regions. When drinking-water resources are degraded – whether by improper treatment and disposal of household sewage, solid waste, overall water management issues or climate-related impacts such as drought – use and consumption of bottled water (which may involve energy-intensive packaging and transport) will likely increase among households that can afford it. The energy required for continued water extraction and purification by public water authorities will also likely grow, further increasing the carbon footprint of water consumption. Similarly, intensive consumption of fresh water resources for sanitation purposes has long-term local and global impacts on the totality of fresh water resource consumption, on availability and quality of local drinking-water supplies, and on the carbon footprint of energy expended for water extraction, purification and sewage treatment.[26]

Water and sanitation issues must be seen in a wider context alongside not only climate change, but other factors that affect water demand and quality.[27] Population growth will cause massive increases in freshwater demand in many parts of Africa and Asia that already face constraints due to the limited infrastructure available to deliver water services.[28] Population growth is also expected to negatively impact water quality as pollution increases, particularly in areas with low sanitation coverage.

Economic growth increases demand for water for all uses, as well as increasing demand for better water supply service and for water-using devices. Such growth also fuels demand for more convenient and potentially water-based sanitation. Urbanization places greater stress on water resources to provide adequate supplies of water within an economically viable distance of settlements, and increases demand due to greater piped water service. Urbanization is also likely to increase pollution.

Human waste, like other forms of organic material, is a source of greenhouse gas emissions. Based on assessments of conventional sewerage and sewage treatment systems, the IPCC estimates that wastewater emits about 590 $MtCO_2$ equivalent of methane, and a further 100 $MtCO_2$ equivalent of nitrogen dioxide. Waste (solid and wastewater combined) would thus account for less than 5% of global emissions.[29] However, greenhouse gas emissions from septic tanks, latrines and open-air defecation remain largely unquantified, IPCC notes, concluding that a more systematic, global assessment is needed.[30,31]

Driven by population growth, urbanization, and overall growth in freshwater demand, climate-change emissions related to wastewater are expected to increase almost 50% by

2020 under business-as-usual scenarios, with largest increases seen in developing countries. Good wastewater management, however, can reduce emissions per unit of waste water produced; so that global sanitation coverage goals might be met without generating an unnecessary climate penalty.[32,33]

Cakir and Stenstrom, for instance, concluded that in the case of low concentration wastewater, aerobic treatment processes release lower levels of greenhouse gases (based on biochemical oxygen demand), but in treatment of highly concentrated wastewater, anaerobic digestion yields lower emissions.[30]

More compact design and planning of residential housing also can reduce the amount of energy, and thus emissions, required for residential wastewater treatment as well as helping to insure access, generating potential health co-benefits in terms of sanitation.

As one simple example, conventional sewers typically operate on gravity, but virtually all require some periodic uplift pumping. If, however, new housing is clustered above the gravity flow lines of major sewage treatment works that would, in turn, reduce energy-demand and may also improve the reliability of service provision. Switching to more energy-efficient pumping technologies, can also reduce greenhouse gas emissions.

Building a closed sewage line over an open waste stream in a new neighborhood on the outskirts of Cochabamba City, Bolivia. (Photo: © Mark Edwards, Hard Rain Picture Library)

A study in South Africa concluded that using well-designed on-site sanitation systems, where possible, produces lower greenhouse gas emissions, per unit of sewage treated, than conventional sewerage systems, due to lower energy requirements, and need not threaten drinking-water supplies.[34] This study also found that recycled water (e.g. grey water reuse) could be used to meet a portion of residential water demand, with a lower carbon footprint than pumping fresh water supplies for all household uses.

While a detailed discussion of mitigation issues related to water supply and sanitation is beyond the scope of this paper, new water conversation and wastewater recycling technologies may thus prove to be cost-effective mechanisms for reducing emissions, as well as a means for assuring long-term resilience of drinking-water supplies essential to health. Carbon footprint analysis of water and sewerage infrastructure can be used as a tool that both reduces climate change emissions and also expands water and sanitation provision at lower energy cost, and thus more affordably.

While the focus of Chapter 1 has been the linkages between housing, environment and climate change, Chapter 2 reviews the specific impacts of the housing environment on major health risks and specific diseases.

References

1. *World Urbanization Prospects: The 2005 Revision*. New York, United Nations, 2006.

2. *CO_2 emissions from fuel combustion: highlights*. Paris, Organisation for Economic Cooperation and Development & International Energy Agency, 2010.

3. *Buildings and climate change: status, challenges and opportunities*. Paris, United Nations Environment Programme, 2007. Rogner H et al. Introduction. In: Metz B et al., eds. *Climate change 2007: mitigation of climate change. Contribution of working group III to the fourth assessment report of the Intergovernmental Panel on Climate Change, 2007*. Cambridge & New York, Cambridge University Press, 2007:95–116.

4. Rogner H et al. Introduction. In: Metz B et al., eds. *Climate change 2007: mitigation of climate change. Contribution of working group III to the fourth assessment report of the Intergovernmental Panel on Climate Change, 2007*. Cambridge & New York, Cambridge University Press, 2007:95–116.

5. Bond T, Venkataraman C, Masera O. Global atmospheric impacts of residential fuels. *Energy for Sustainable Development*, 2004, 8(3):20–32.

6. MacCarty N et al. A laboratory comparison of the global warming impact of five major types of biomass cooking stoves. *Energy for Sustainable Development*, 2008, 12(2):5–14.

7. UNEP/WMO. Integrated assessment of black carbon and tropospheric ozone: summary for decision makers. Nairobi, UNON/Publishing Services Section, 2011.

8. Ramanathan V, Carmichael G. Global and regional climate changes due to black carbon. *Nature Geoscience*, 2008, 1(4):221–227.

9. The World Resources Institute. (http://www.wri.org)

10. Levine M et al. Residential and commercial buildings. In: Metz B et al., eds. *Climate change 2007: mitigation of climate change. Contribution of working group III to the fourth assessment report of the Intergovernmental Panel on Climate Change, 2007*. Cambridge & New York, Cambridge University Press, 2007:387–446.

11. Legros G et al. *The energy access situation in developing countries: a review focusing on the least developed countries and Sub-Saharan Africa*. New York, World Health Organization & United Nations Development Programme, 2009.

12. *Global health risks: mortality and burden of disease attributable to selected major risks*. Geneva, World Health Organization, 2009.

13. Adair-Rohani H, Bruce N. *Health in the green economy: The household energy sector in developing countries*. Geneva, World Health Organization, 2011.

14. Glaeser E, Kahn M. *The greenness of cities: carbon dioxide emissions and urban development*. Cambridge, National Bureau of Economic Research, 2008.

15. Norman J, MacLean H, Kennedy CA. Comparing high and low residential density: life-cycle analysis of energy use and greenhouse gas emissions. *Journal of Urban Planning and Development*, 2006, 132(1):10–21.

16. *Why urban health matters: World Health Day 2010*. Geneva, World Health Organization, 2010.

17. *Hidden cities: unmasking and overcoming health inequities in urban settings*. Kobe, World Health Organization / WHO Centre for Health Development & United Nations Human Settlements Programme, 2010.

18. Kinyanjui M et al. Development context and the millennium agenda. In: *The challenge of slums: global report on human settlements 2003, revised and updated version (April 2010)*. New York, United Nations Human Settlement Programme, 2010.

19. *State of the world's cities 2006/7*. Nairobi, Kenya, United Nations Human Settlements Programme, 2006.

20. *Climate change and its health implications: a summary report for environmental health practitioners on the health implications of climate change*. London, Chartered Institute of Environmental Health, 2008.

21. Oke TR. City size and urban heat island. *Atmospheric Environment*, 1973, 7(8):769–779

22. Wilby RL. Past and projected trends in London's urban heat island. *Weather*, 2003, 58:251–260.

23. *Protecting health from climate change: World Health Day 2008*. Geneva, World Health Organization, 2008.

24. *WHO guidelines for indoor air quality: dampness and mould*. Geneva, World Health Organization, 2009.

25. *Climate change and health*. Fact sheet No. 266. Geneva, World Health Organization, 2010. (http://www.who.int/mediacentre/factsheets/fs266/en/index.html)

26. Jordan, water is life. In: *Health and environment: managing the linkages for sustainable development, a toolkit for decision-makers*. Geneva, World Health Organization, 2008:50–57.

27. Howard G, Bartram J. *Vision 2030: The resilience of water supply and sanitation in the face of climate change*. Geneva, World Health Organization, 2010.

28. *Water for food, water for life: a comprehensive assessment of water management in agriculture*. London, Earthscan & Colombo, International Water Management Institute, 2007.

29. Bogner J et al. Waste management. In: Metz B et al., eds. *Climate change 2007: mitigation of climate change. Contribution of working group III to the fourth assessment report of the Intergovernmental Panel on Climate Change, 2007*. Cambridge & New York, Cambridge University Press, 2007:585–618.

30. Cakir FY, Stenstrom MK. Greenhouse gas production: a comparison between aerobic and anaerobic wastewater treatment technology. *Water Research*, 2005, 39:4197–4203.

31. Bates BC et al., eds. *Climate change and water*. IPCC Technical Paper Series, No. 6. Geneva, Intergovernmental Panel on Climate Change Secretariat, 2008.

32. El-Fadel M, Massoud M. Methane emissions from wastewater management. *Environmental Pollution*, 2001, 114:177–185.

33. Prendez M, Lara-Gonzalez S. Application of strategies for sanitation management in wastewater treatment plants in order to control/reduce greenhouse gas emissions. *Journal of Environmental Management*, 2008, 88:658–664.

34. Freidrich E, Pillay S, Buckley CA. Carbon footprint analysis for increasing water supply and sanitation in South Africa: a case study. *Journal of Cleaner Production*, 2009, 17:1–12.

Notes

2

Bratislava, Slovakia: The absence of balconies can limit access to daylighting, and a dearth of green space limits opportunities for outdoor activity and social interactions. Block-style construction is often a source of water leakage.

Matthias Braubach

Review of housing and health risks

2.1 A framework for understanding health risks in housing

WHO defines housing in terms of four related dimensions:

- The house (or dwelling) is the physical structure used or intended to be used for human habitation.
- The 'home' is the economic, social and cultural structure established by the household.
- The neighbourhood is the immediate environment, including adjacent housing. areas, streets, shops, places of worship, recreational and green spaces, and transport.
- The community includes those who live, work and provide services in the neighbourhood.[1]

Within this context, housing and health risks are often defined very broadly in terms of physiological risks, psychological risks, risks of infection and risks of injury.[2,i] In all cases, the association between housing and health is complex, and causal relationships can be hidden in or otherwise influenced by confounding variables and effect modifiers.[3] Climate change mitigation strategies can directly and indirectly affect housing and health. Presenting the full range of housing and health impacts allows systematic consideration of climate change mitigation policies that reduce health risks and generate optimal health co-benefits.

This chapter summarizes key environmental risks to health in the housing environment, as well as specific health impacts in term of diseases and injuries. In cases where one risk is a major factor in multiple diseases, e.g. indoor air pollution, links to health are described in terms of the risk. When a disease (e.g. TB) may be transmitted by more than one housing risk factor, this is discussed. While acknowledging the inevitable overlap, the complementary categories help clarify the issues.

---

[i] For example, China defines healthy housing "on the basis of meeting basic elements of the housing construction, improving and preserving the physiological, psychological, moral and social health, and promoting the sustainable development of the housing constructions to further improve the housing quality and the residential environment." Source: China National Engineering Research Centre for Human Settlements. *Technical specification for construction of healthy housing*. Beijing, China Planning Press, 2009.

Housing-related environmental health risks

- Indoor air quality risks (indoor smoke, mould, radon, chemicals, asbestos and lead)
- Extreme thermal conditions
- Pests and infestations
- Noise
- Urban form and density

Housing-related diseases and injuries

- TB and other airborne infectious diseases
- Vector-borne diseases
- Waterborne diseases
- Domestic injuries
- Mental health

2.2 Environmental health risks

2.2.1 Indoor air quality risks

Indoor smoke from household heating/cooking

Indoor smoke from solid fuel combustion is the eighth most important risk factor in burden of disease and is responsible for 2.7% of the global burden of disability-adjusted life years (DALYs; Fig. 5).[4] In 2004, indoor air pollution from solid fuel use was responsible for almost 2 million deaths,[4] making this risk factor the second major environmental contributor to ill health (behind unsafe water and sanitation). Indoor air pollution particularly affects children and women, who spend more time at home and are in closer proximity to the flame. Incomplete combustion of coal and biomass fuels in inefficient traditional cookstoves exposes household members to high levels of health-damaging pollutants; these stoves are used by 3 billion people, mostly in developing countries. In high-mortality developing countries, indoor smoke from combustion sources is responsible for an estimated 3.7% of the overall disease burden, making it the most dangerous killer after malnutrition, unsafe sex and lack of safe water and sanitation.[4]

Indoor smoke has been associated with a wide range of health outcomes, with a high risk of acute lower respiratory illness, commonly pneumonia. Nearly half of childhood pneumonia is attributed to indoor smoke from household heating in developing countries.

COPD (chronic obstructive pulmonary disease) due to indoor smoke from inefficient biomass and coal stoves causes over 1 million premature deaths each year, primarily among women in poor countries. A significant proportion of ischaemic heart disease and lung cancer is also due to use of such stoves. Lung cancer risks among those exposed to indoor smoke from coal are 1.5 times over the average for men and 1.9 times more for women, and including about 36 000 premature deaths annually.[6] Other health outcomes associated with indoor air pollution include; low birthweight and perinatal mortality (stillbirths and deaths in the first week of life), asthma, otitis media (middle ear infection), other acute upper respiratory infections, tuberculosis (See 2.3.1), nasopharyngeal

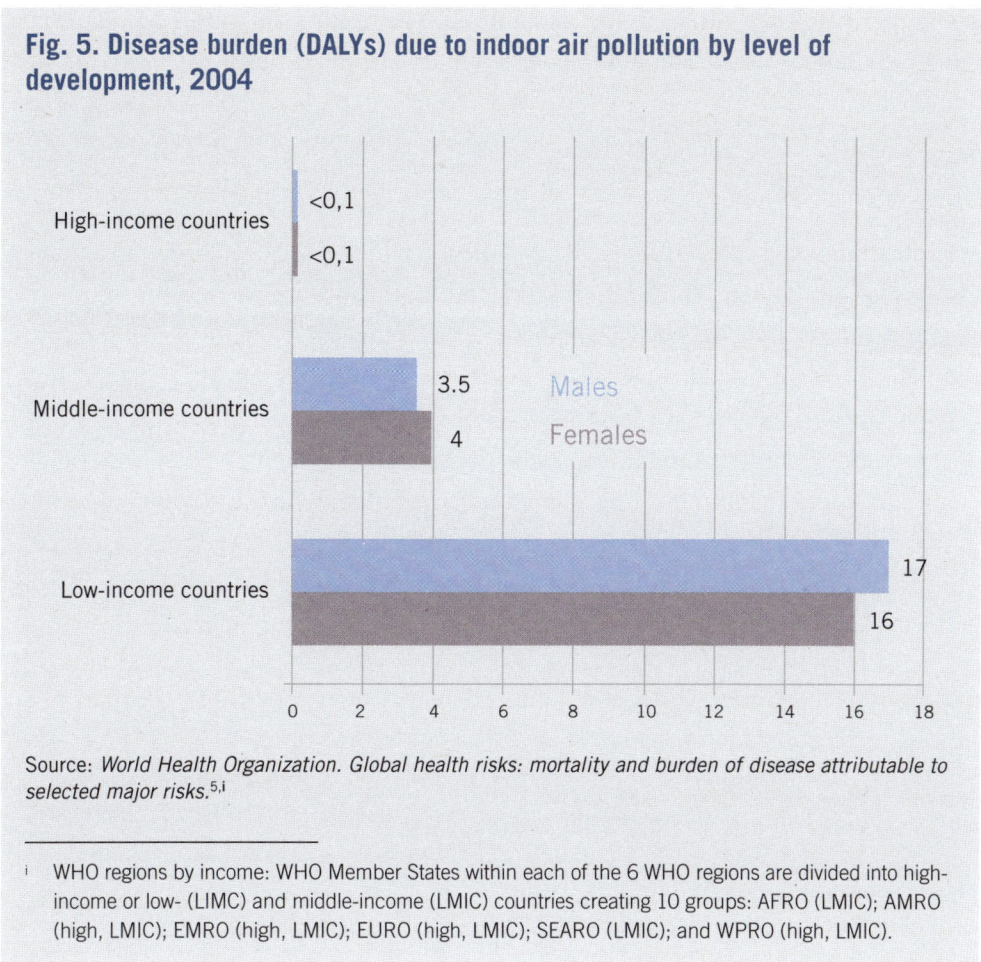

Fig. 5. Disease burden (DALYs) due to indoor air pollution by level of development, 2004

Source: *World Health Organization. Global health risks: mortality and burden of disease attributable to selected major risks.*[5,i]

---

[i] WHO regions by income: WHO Member States within each of the 6 WHO regions are divided into high-income or low- (LIMC) and middle-income (LMIC) countries creating 10 groups: AFRO (LMIC); AMRO (high, LMIC); EMRO (high, LMIC); EURO (high, LMIC); SEARO (LMIC); and WPRO (high, LMIC).

and laryngeal cancer, cataract (blindness) and cardiovascular disease. Further discussion of the health impacts of biomass and coal-burning systems is provied in the companion *Health in Green Economy* report previously referenced.

Infiltration of outdoor air pollution indoors is also a health risk, and for several pollutants this may be the major pathway of indoor exposures. Finally, tobacco smoke is a major source of health-damaging indoor air pollution, common in the housing environment. Tobacco smoke contains measurable quantities of carbon monoxide, ammonia, nicotine, hydrogen cyanide, particulate matter (PM) and a number of carcinogens.[7] Many of the toxic chemicals present in tobacco smoke also are present in wood smoke, creating a double burden for those exposed to both hazards.

There are more than 1.3 billion smokers worldwide, with around 82% residing in low- and middle-income countries.[8] Smoking has been identified as the leading cause of preventable disease and premature death in industrialized countries. By 2030, a projected 8 million people in developing countries will be killed by tobacco every year.[9]

The use of incense in many countries as a cultural ritual can also be considered a health risk. Incense is composed of aromatic biotic materials that release fragrant smoke when burnt. It has been documented that incense use is associated with a cluster of neurological symptoms, including headache, dizziness, difficulty in concentration and forgetfulness.[10–12]

Cooking and heating with unvented gas and kerosene heaters can cause potentially fatal acute poisoning from carbon monoxide as well as chronic exposure to other combustion pollutants.[13]

Mould and moisture

Mould on residential ceiling and wall. (Photo: Juergen Rath)

Microbial pollutants and allergens (e.g. pollen, bacteria, fungi, microbes) are identified as relevant indoor exposures exceeding outdoor concentrations. Active growth of microorganisms and excessive accumulation of biological agents in the indoor environment is often due to dampness and inadequate ventilation as well as inappropriate occupant behaviour. Excess moisture on almost all indoor materials leads to growth of mould, fungi and bacteria; these emit spores, cells, fragments and volatile organic compounds into indoor air. Moreover, dampness initiates chemical or biological degradation of materials that also pollutes indoor air. Dampness is therefore considered to be a strong, consistent indicator of risk of asthma and respiratory symptoms (e.g. cough and wheeze).[14, 15] Moisture control is therefore an essential element for healthy housing.

Dampness is more likely to occur in houses that are overcrowded and lack appropriate heating, ventilation and insulation.[16]

Radon, VOCs and other chemical pollutants

The trend towards making buildings more airtight for energy efficiency also increases risks of indoor pollutant accumulations, making pollution mitigation measures all the more important. Chemical and biological contaminants may be reduced by effective source control, moisture control, ventilation and higher overall rates of air exchange.

Venting the soil under and around a home helps prevent radon gas from seeping into floors. (Photo: Cincinnati Habitat for Humanity: http://cincinnatihabitat.org)

Radon is a radioactive gas that emanates from rocks and soils and tends to concentrate in enclosed spaces like underground mines or houses. Soil gas infiltration is recognized as the most important source of residential radon. Other sources, including building materials and water extracted from wells, are of less importance in most circumstances. Radon is a major contributor to the ionizing radiation dose received by the general population.

Recent studies on indoor radon in Europe, North America and Asia provide strong evidence that radon causes a substantial proportion of lung cancers in the general population. Current estimates of lung cancers attributable to radon range from 3–14%, depending on the average radon concentration in the country concerned and calculation methods.[17]

Evacuation of radon gases usually requires special design measures (sub-slab depressurization). Addressing radon is important both in new construction (prevention) and in existing buildings (mitigation or remediation). The primary radon prevention and mitigation strategies focus on reversing air-pressure differences between the indoor occupied space and the outdoor soil to prevent radon from entering and concentrating indoors. Sealing radon entry routes can help in this effort but is not considered a stand-alone method. In many cases, a combination of strategies best reduces radon concentrations.

A range of other indoor air pollutants, including volatile organic compounds (VOC) such as formaldehyde,[ii] also may contribute to morbidity and mortality, including acute conditions (e.g. poisonings) as well as cancers and other chronic conditions. There is consistent evidence linking asthma and allergy incidence to a range of indoor chemical pollutant exposures.

Indoor pollutants often are emitted slowly from indoor building materials, furniture, paints and carpets. They also may be released by human activities (e.g. smoking and detergent use), or drift in from attached garages or other outdoor sources.[18] Flourescent and compact flourescent lights as well as thermostats commonly contain mercury and thus must be handled with care to avoid exposures, particularly when broken (see Chapter 3.9). Risks of indoor air pollutants can be lowered by adequate natural ventilation, but also through the use of healthier building materials, including replacement or phasing out of hazardous building substances wherever possible.

Asbestos and other natural mineral fibres

Asbestos fibres may be another source of indoor air pollution. Due to extraordinary tensile strength, poor heat conduction and relative resistance, asbestos has been widely used for sprayed fire protection, thermal and acoustic insulation, pipe and boiler lagging, ceiling tiles, partitioning, roofing and cladding. Asbestos also is used in roofing shingles, water supply lines, asbestos cement, patching and joint compounds. Where these materials are damaged or deteriorate, they can release dust and fibres that become airborne. Chrysotile (white asbestos) is the form of asbestos most widely used, together with amosite (brown asbestos) and crocidolite (blue asbestos).

All forms of asbestos are carcinogenic to humans, and may cause mesothelioma and cancer of the lung, larynx and ovary when inhaled. Asbestos exposure is also responsible for other diseases, such as asbestosis (fibrosis of the lungs), pleural plaques, thickening and effusions. These pathologies have a latency period of 20 to 30 years. Fibrous mineral silicates are naturally occurring, and along with asbestos include minerals such as vermiculite.

Typical asbestos cement roofing in a slum in India. Note the lack of indoor ventilation, apart from that available by the doorway. (Photo: Nick Clarke)

Currently, about 125 million people worldwide are exposed to workplace asbestos, and one in every three deaths from occupational cancer is estimated to be caused by asbestos. According to the most recent WHO estimates, more than 107 000 people die each year from asbestos-related lung cancer, mesothelioma and asbestosis resulting from exposure at work. In addition, it is estimated that several thousand deaths annually can be attributed to exposure to asbestos in the home.[19] In the industrialized countries of western Europe, North America, Japan and Australia, 20 000 asbestos-induced lung cancers and 10 000 mesothelioma cases are estimated to occur annually.[20]

Due to these cancer and mesothelioma risks, the use of asbestos, including as a housing insulation material, has been banned in more than 50 countries and replaced by safer substitutes. Most developed countries strictly regulate asbestos control and management during building retrofits and demolition. Nonetheless, asbestos is still being mined and manufactured, and it is widely used in the housing sector in many developing countries.

---

ii *Formaldehyde.* Lyon, International Agency for Research on Cancer / World Health Organization. (IARC Monographs on the evaluation of carcinogenic risks to humans, volume 88) (http://monographs.iarc.fr/ENG/Monographs/vol88/mono88-6.pdf)

This results in potentially large but unknown environmental exposures, not only during asbestos mining and manufacture, but also in housing construction and demolition. During the clean-up of damaged and destroyed buildings after natural disasters such as tsunamis and earthquakes, particularly in developing countries, the risk of exposure to asbestos increases among volunteers and local residents. These groups may be unaware of asbestos hazards or unable to identify asbestos-containing materials. Weathering and the ageing process may also lead to degradation of the asbestos fibres, with increased risk of human exposure.

Lead

Old paint can contain lead, which is toxic to humans and especially dangerous to children. (Photo: istockphoto)

Even low levels of lead dust exposure are associated with impairment of childhood cognitive function and abnormal infant behaviour. Within the housing environment, lead is commonly found in lead-based paints and lead-containing water pipes; both are possible sources of exposure.[21] Over the past 30 years regulatory and environmental reforms have phased out leaded gasoline in all but a handful of countries worldwide. Developed countries have also outlawed use of lead paints, further lowering risks of child exposures. In many developing countries, however, lead paint and products remain in the market, and regulations for testing and removal of lead from housing paint and pipe products are not well established. In the absence of routine testing, the prevalence of lead in housing environments is not well understood.

Lead is stored especially in the bones, but also in the blood, liver and kidneys, where it has toxic effects. Long-term exposure to lead mainly affects the nervous system. Due to their low body weight and developing nervous systems, children are the most vulnerable. Young children can suffer declines in intellectual performance; with increasing blood lead concentrations, cognitive functioning decreases and neurological and developmental deficiencies increase. Also, lead in paint or dust can be easily ingested by small children crawling on the floors of exposed homes. In homes with lead exposure, mitigation is often possible only by removing or covering building components coated with lead-based paint.[22]

2.2.2 Extreme thermal conditions

Indoor thermal conditions are a major determinant of cardiovascular and respiratory health problems. Often thermal conditions are linked to poorly insulated, poorly heated or poorly ventilated buildings, but in many instances such problems have a social dimension, as poor households may not be able to afford to heat their homes adequately. There are as yet no global figures for mortality and morbidity from excessive heat and cold, although in 2011 WHO plans to publish global burden of disease estimates of heat- and cold-related deaths associated with climate change.

At the same time, a 2008 systematic review of 16 studies examining excess winter mortality, socioeconomic status and housing quality found that "studies to date do not provide good evidence that housing quality or socioeconomic status affect excess winter mortality and excess winter hospitalization … the evidence of linkage is inconsistent, with some studies showing a weak protective effect of home heating."[23]

Notably, such studies have mostly been conducted in developed countries, and along with housing quality or socioeconomic status as such, a wide range of other potential confounders exist within and between populations of different countries and regions, including ethnic and physiological differences, winter behaviours, clothing, time spent outdoors and health status generally. These are in addition to housing factors such as crowding, sunlight exposure and indoor air pollution.[23]

Even so, many European countries have begun estimating deaths related to excess cold and heat waves. The United Kingdom estimates that between 2004 and 2008, more than 130 000 people over 65 died from cold-related illnesses during the winter months. Similar cases of excess winter mortality are observed in regions such as the Baltic countries, eastern Europe and central Asia.[24] The elderly are the predominant victims of heat and cold waves.

In the United Kingdom, five main building-related determinants of cold indoor air temperatures have been identified as:[25]

- Age of the dwelling (the older, the colder)
- Absence of / dissatisfaction with the heating system
- Cost of heating (highest is colder)
- Low household income (less is colder)
- Household size (smaller is colder)

On the other hand, extreme heat exposure can lead to exhaustion, heat cramps, heat stroke and ultimately death, largely from cardiovascular disease. Heat waves characterized by long duration and high intensity have the highest impact on mortality.[26] Nearly 45 000 excess deaths were observed in 12 European countries in August 2003 during a prolonged heat wave.[27] WHO's Euro-heat project showed that the impact of heat waves characterized by longer duration (more than four days) was 1.5–5 times higher than for short heat waves.[28] Studies also show an association between such mortality and social isolation of the elderly population, and between heat-related mortality and housing conditions (Fig. 6).

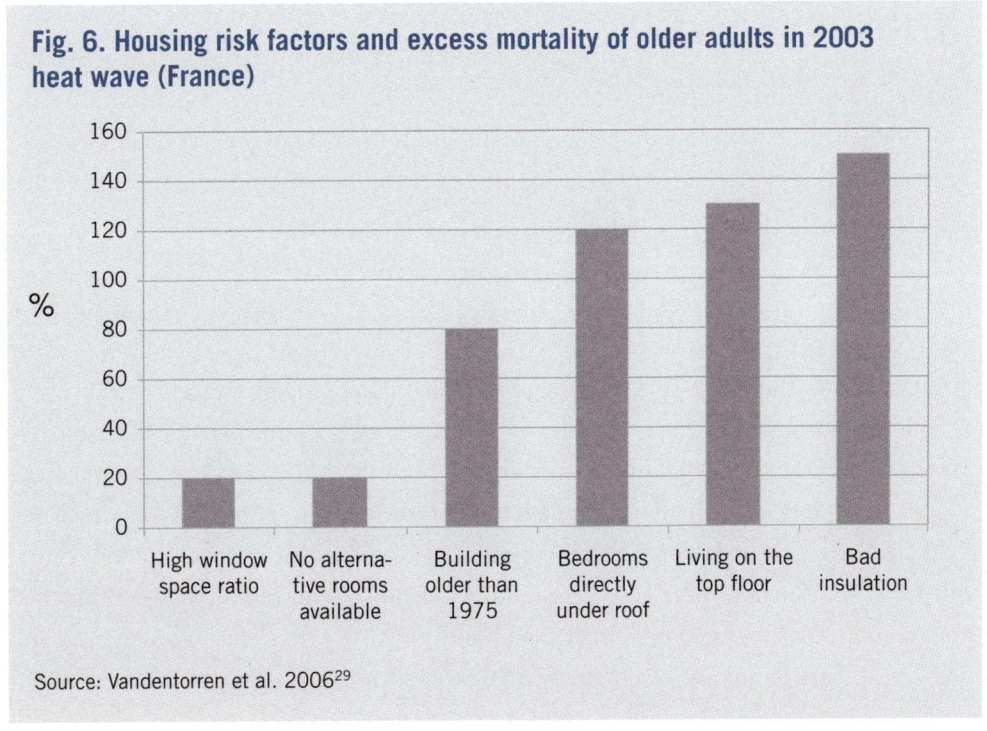

Fig. 6. Housing risk factors and excess mortality of older adults in 2003 heat wave (France)

Source: Vandentorren et al. 2006[29]

2.2.3 Pests and infestations

Urban sprawl, irresponsibly discarded trash, international travel and climate change are all factors caused by human behaviour that increase population exposures to pests and pest-related diseases. Increasing environmental change brings new risks from pests and the diseases they are associated with. As cities expand in peri-urban areas and in new rural settlements on the edges of forests and jungles, people in temperate zones are more exposed to vector-borne diseases, including tick-borne diseases such as Lyme disease and tick-borne encephalitis in temperate regions, and in Latin America, diseases such as Chagas and leishmaniasis. These severely disabling diseases have spread over the past 30 years, in part because of human actions, despite pest management techniques now available.[1] Simple maintenance failures such as broken roof tiles, damaged water pipes and overflowing cisterns, together with common mistakes in design or construction such as excessive use of impermeable membranes, can lead to home infestation by harmful insects and pests. Several studies have suggested that exposure to cockroach allergens is one of the most important risk factors for asthma in inner-city or congested households.[30-34]

In accordance with different categories of pests, there are also different health outcomes. In Latin America, Chagas disease has one of the largest disease burdens in terms of morbidity.[35] The burden of disease from leishmaniasis in Latin America as well as more severe visceral leishmaniasis (VL) in South-East Asia remains severe. Common pests found in temperate zones can trigger allergic reactions among sensitive people through exposure to the animals' body particles, excretions and emissions. House dust mites and cockroaches are among the most relevant pest triggers. It has been estimated that 10–20% of the population is potentially allergic to dust mites, while up to 70% of asthmatic people are likely to show allergic reactions.[1]

Excessive moisture in buildings leads to house dust mites. Studies have found that relative humidity is higher in dwellings where the ground floor consists of a concrete slab in direct contact with the ground. If the floor covering is absorbent – a carpet, for example – it can act as a reservoir leading to long-term dampness. Ventilation creates conditions that kill mites in cold winter; ventilation also reduces exposure to mite allergens and other indoor airborne pollutants. Although the health benefits of insulation are similarly obvious, its effects on mite populations are not so clear-cut. Modelling studies suggest that the promotion of mite growth from rising room temperatures tends to be outweighed by falling relative humidity.

2.2.4 Noise

Research shows that excessive noise levels result in sleep disturbances, cardiovascular and psycho-physiological problems, performance reduction, increased annoyance responses, adverse social behaviour and, at very high levels, hearing loss. A significant amount of literature confirms that noise poses a serious threat to children's hearing, health, learning and behavior.[36-41] Environmental noise is the primary exogenous cause, with traffic noise usually leading neighbourhood noise and aircraft noise; other factors are high density, poor housing quality and housing proximity to commerce and industry.[1] Environmental noise acts as a stressor by disturbing sleep and represents an annoyance during the day.[42] Sleep is essential for humans, and acute sleep disturbances affect qualitative and quantitative performance and increase the risk of domestic accidents.

A maximum noise level of 30dB(A) has been recommended for bedrooms to prevent sleep disturbance, and a limit of 35dB(A) for the indoors of dwellings more generally.[43]

2.2.5 Urban form and density

While there is no "ideal size" for urban settlements,[44] residential development in geographically sprawling and isolated areas will tend to limit independent mobility of children, the elderly and women and disabled, who generally have less access to travel by car. This, in turn, may limit access to employment, education, leisure outlets, primary health care, fresh food stores and commerce and other community facilities.

Links between neighbourhood design and physical activity have become a focus of recent health research, which finds current obesity trends closely linked to urban designs favouring car transport.[45] Heavily trafficked urban or suburban residential neighborhoods may have comparatively less space for physical exercise (sport facilities, playgrounds, parks and other open spaces) as well as little infrastructue for active travel, e.g. networks of sidewalks and bike tracks well connected to shopping and employment and pleasant to use. Recent research has demonstrated strong associations between the physical environment and residents' physical activity patterns, as well as levels of physical fitness and body mass index (BMI).[46]

> Housing design and neighborhood density exert an influence on children's development. In some settings, mixed use, medium density neighborhoods make it easier for children to move independently, getting exercise and developing motor and cognitive skills.

Housing design and neighbourhood density also exert a powerful influence on children's independent mobility. For instance, mixed-use, medium-density neighborhoods may facilitate routine physical activity if children can move safely and independently on foot or bicycle to school friends and shops. Conversely, high-rise blocks with many stairs or an elevator and streets with no sidewalks may create barriers for children. Along with physical activity, the ability to move about freely can help a child practice certain motor and cognitive skills important to his/her development.[47,48] Design and accessibility features also impact abilities of older and/or disabled persons to live independently in their own homes and to make complete use of their dwellings and their immediate environment, regardless of age or physical condition.[49,50]

Research also indicates that residents' perceptions of and satisfaction with their residential environmental quality are closely linked to factors such as social ties in the neighbourhood, safety risks (crime, traffic), environmental hygiene (noise, air pollution) and the presence of facilities (shops, greenery).[51–56]

Slums

Slum housing environments exacerbate many of the same housing and health risks described in this chapter, as people are simultaneously confronted with low incomes, limited educations, insufficient diets, overcrowding, pollution, stress, traffic, social instability and insecurity. Economic stress prompts many slum dwellers, including children, to take jobs in hazardous conditions exposing them to higher occupational health risks.

The residents of this Mexico City settlement, Borde de Vias Colony, moved here from the countryside when their land eroded and became unproductive. Some 1000 rural migrants a day are estimated to arrive in the city. (Photo: © Mark Edwards, Hard Rain Picture Library)

And combined social and economic stresses may foster increased alcohol and drug abuse, as well as greater risks of exposure to sexually transmitted diseases.

Studies indicate that the high prevalence of pneumonia, diarrhoea, malaria, measles, TB and other major infectious diseases, however, is often due to poor living conditions rather than income levels per se. For instance, municipal supplies of safe drinking-water rarely connect to slums, and it is common for slums' pit latrines to be shared by thousands of people. Children from slums have higher rates of diarrhoea than children of the poorest rural families because they are exposed to contaminated water and food.[57]

2.3 Diseases and injuries

2.3.1 TB and other airborne infectious diseases

One of the first associations between housing and health was noted by tuberculosis (TB) researchers. The typhoid and tuberculosis experiences in the 19th and early 20th centuries showed that basic improvements in sanitation, ventilation, reduced household crowding and other housing improvements helped conquer these epidemics.[3]

As TB is once more a global epidemic, higher rates of person-to-person contact in dense urban settlements spread the infection. An estimated 1.3 million people died from TB in 2008. The highest number of TB deaths was in South-East Asia, while the highest TB mortality per capita was in Africa.[58]

In the case of TB, health literature documents a significant association between housing density, isolation, income levels and TB. Overcrowded housing has the potential to increase exposure of susceptible individuals to infectious TB cases, and isolation from health services may increase the likelihood of TB.[59] Lack of ventilation and lack of light favour the proliferation and transmission of mycobacteria.

Along with crowding and density, the literature also notes linkages between TB and exposures to indoor smoke from the burning of solid fuels, and particularly from use of kerosene lighting. In one recent study in Nepal of 125 women hospitalized for TB, a higher proportion used kerosene lighting than a TB-negative control group. This difference was statistically significant even when other factors such as crowding, socio-economic level and housing conditions were considered in the analysis.[60]

Similar biological mechanisms and pathways facilitate transmission of other infectious diseases in crowded households. These are summarized by the United Nations Commission on Human Settlements, which notes that high household population density and occupancy conditions increase risks of infection generally in terms of the risk of multiple infections, severity of infection, risk of long-term negative impacts of infections and risk of disease transmission due to increased proximity of people, as measured in numbers of persons per habitable room, floor area or bed.[61–63]

Finally, the number of air exchanges per hour in a room will directly affect the risks of disease transmission: the risk lessens when there are more air exchanges. These quantitative associations have been identified in WHO work on natural ventilation in health care settings, and are described in more detail below.

2.3.2 Vector-borne diseases

Climate change is likely to cause changes in ecological systems that will affect the risk of infectious diseases in Europe, including seasonal activity of local vectors and invasion of tropical and semi-tropical species into areas where they previously could not survive due to temperature and environmental conditions. Shifts in global and regional distribution and behaviour of insect and bird species are early signs that biological systems are already responding to climate change. Patterns of infectious disease are and will be affected by the movement of people and goods, and by changes in hosts and pathogens, land use and other environmental factors. Personal risk factors such as immune system status also play important roles.

Densely populated urban areas may become increasingly vulnerable to vector-borne diseases due to climate change, as shifting climate patterns extend the range of certain vectors. Rapid unplanned urbanization can produce mosquito breeding sites, high human population densities provide a large pool of susceptible individuals, and increased temperatures increase absolute humidity that can also extend species range.[64] Diseases spread in this way include dengue fever,[65-67] malaria, Chagas and filariasis. Although climate change is likely to result in the expansion of malaria-carrying mosquitoes to some new locations, it is likely to cause the contraction of this range in other places.[68]

In the context of preventing vector-borne diseases, housing plays an important role. This relationship may be direct or indirect. The design of the housing structure helps determine what opportunities certain vector populations have to breed and circulate in proximity to humans, and the ease with which they may come into sufficiently close and frequent contact to transmit disease.

Housing design can impede vector entry; mosquitoes cannot enter easily through well-maintained window screens or closed windows. Cracks in walls or use of certain construction materials can facilitate vector habitats. Triatomine bugs typically live in the walls and cracks of poor homes in Latin America, as do sand flies (carrying visceral leishmaniasis) in homes of the Indian subcontinent. Human behaviour in and around the house may determine vector habitats: for instance, bednets provide protection against mosquito bites, protection of water storage containers can impede breeding of dengue vectors in endemic areas, and housing in close proximity to domestic animals may expose humans to vectors. There are thus clear relationships between poor housing conditions and vector-borne diseases, just as there are clear relationships between poverty and vector-borne diseases.

This does not imply that high-cost housing is necessary to protect against vectors and vector-borne diseases, or that at-risk human behaviour is the sole contributor to vector-borne diseases. To the contrary, fairly cheap and effective interventions may be possible if these are accepted and used by the people and communities affected.

In the context of preventing vector-borne disease, housing plays an important role. The housing structure helps determines what opportunities certain vectors have to breed and circulate near humans, and the ease with which they may be able to transmit disease.

2.3.3 Waterborne diseases related to unsafe water and sanitation

Access to clean water and sanitation are foundations of a healthy housing environment. Around 884 million people globally do not use an improved source of drinking-water; 2.6 billion people do not have access to any type of improved sanitation facility.[69] And about 2 million people die every year due to primarily water and sanitation-related diarrhoeal diseases, the majority children less than 5 years of age. The most vulnerable populations in developing countries are normally peri-urban or rural inhabitants.[69]

Both floods and droughts occuring as a result of climate change may threaten drinking-water quality and sanitation. Increasing drought and water shortages, as well as increased flooding due to warming trends and severe weather, impact upon piped water and sanitation. Flooding can rapidly disperse faecal contaminants and contaminate water and food, leading to outbreaks of waterborne diseases such as cholera and diarrhoeal diseases; droughts can result in acute water shortages, leading to more diarrhoeal disease.[70]

Both increased flooding and increased droughts as a result of climate change impact on water and sanitation risks. Providing facilities at household level for sanitary disposal of excreta, safe water pipes and storage, and good hygiene all are important.

Providing access to sufficient quantities of safe drinking-water and sanitation in this context is therefore a challenge. At household level, the provision of facilities for good hygiene, sanitary disposal of excreta, the use of safe materials for water pipes, safe water storage, and knowledge about sound hygiene behaviours are critical to reducing the immediate burden of waterborne diseases. As discussed in Section 1.5.5, more energy-efficient provision of water and waste water infrastructure, along with greater reliance upon household water conservation measures (e.g. rainwater harvesting, onsite grey water reuse) may become increasingly important to resilience, and to achieving or maintaining long-term reductions in the burden of waterborne diseases.

2.3.4 Domestic injuries

In 2004, injuries represented approximately 15% of the years of life lost (YLL) per 1000 population.[71] While traffic injury is the overwhelming cause, unintentional home injuries are also a serious public health problem. There are no global statistics regarding domestic injuries; however, regional data and case study evidence do exist. In the years 2002–2004, home injuries were the leading cause of injury death in children under 5 years of age in 16 European countries.[72] In the United Kingdom in 1999, there were 2.8 million home accidents requiring medical attention for an estimated cost of around 35.5 billion euros.[73]

A WHO study of four low-income countries found that 65% of childhood burns as well as most falls, drowning and poisonings occur in and around the home.[74] Domestic accidents include (not in order of occurrence) cuts, falls, drowning, collisions, choking, burns, electrical shock and poisoning. Along with human behaviour, however, design and maintenance of housing is a leading cause of domestic injury.[75,1] This can include poor design of gas and electrical installations, steps and stairs, and windows and balcony features on upper floors. Slippery floor materials, poor lighting, noise exposure and crowding also increase accidents. WHO's LARES study showed a very strong correlation between accidents in the home and housing conditions.[75,1] A New Zealand study found injuries that occurred in the home were the most common cause of hospitalization due to injury, and the more hazards found in the home, the more injuries occurred.[76,77]

2.3.5 Mental health

One of the primary functions of housing is to provide shelter from outside aggression. Beyond that function, however, a dwelling is defined as a protective physical and psychological envelope for individuals and for the intimate dimension of familial or social relationships. Poor-quality housing that provides insufficient protection from the outside, from scrutiny and from intrusion, can be a source of suffering. Such events may generate pathological manifestations such as anxiety, depression, insomnia, paranoid feelings and social dysfunction.[78] Several studies have highlighted the influence of environmental factors such as pollution, noise and crowding on mental health, depression symptoms and social well-being.[79] A recent review of epidemiological surveys in Europe associated a pattern of poorer mental health with high-density multi-unit dwellings, although the quality of research reviewed was poor and may be confounded by multiple cultural and socioeconomic factors.[80] Stressful housing conditions can aggravate pre-existing psychiatric pathologies.[1] Prolonged heat and cold events can create stress situations that may initiate or exacerbate health problems in populations already suffering from mental disease and stress-related disorders.[81]

Sudden catastrophic events also have a variety of impacts on mental health, increasing short-term emotional stress as well as anxiety about the future. These, too, may become more pronounced as a result of more climate change-related extreme weather and natural disasters, such as hurricanes, wildfires and floods.

References

1. Bonnefoy X. Inadequate housing and health: an overview. *International Journal of Environment and Pollution*, 2007, 30(3/4):411–429.
2. *Housing health and safety rating system: operating guidance.* London, Office of the Deputy Prime Minister, 2006.
3. Jacobs DE et al. A systematic review of housing interventions and health: introduction, methods, and summary findings. *Journal of Public Health Management Practice*, 2010, 16(5):S5–S10.
4. *Indoor air pollution: global burden of disease due to indoor air pollution.* Geneva, World Health Organization, 2004. (http://www.who.int/indoorair/health_impacts/burden_global/en/index.html)
5. *Global health risks: mortality and burden of disease attributable to selected major risks.* Geneva, World Health Organization, 2009.
6. Ezzati M et al., eds. *Comparative quantification of health risks: global and regional burdens of disease attributable to selected risk factors.* Geneva, World Health Organization, 2004.
7. Passive Smoking. In: *The health consequences of smoking – A report of the Surgeon General.* Washington, Department of Health and Human Services, 1984.
8. *Tobacco Addiction.* Geneva, WHO Disease Control Priorities Project, April 2006.
9. *WHO Report on the Global Tobacco Epidemic: The MPOWER Package.* Geneva, World Health Organization, 2008.
10. Tong Szeto Y et al. Effects of incense smoke on human lymphocyte DNA. *Journal of Toxicology and Environmental Health*, 2009, Part A, 72(6):369–373.
11. Al-Rawas O, Al-Maniri A, Al-Riyami B. Home exposure to Arabian incense (bakhour) and asthma symptoms in children: a community survey in two regions in Oman. *BMC Pulmonary Medicine,* 2009, 9:23.
12. Siao WS, Balasubramanian R, Joshi UM. Physical characteristics of nanoparticles emitted from incense smoke. *Science and Technology of Advanced Materials*, 2007, 8:25–32.
13. *Protect your family and yourself from carbon monoxide poisoning.* Washington, US Envrionmental Protection Agency, Indoor Environments Division, Office of Air and Radiation, 1996.
14. Fisk WJ, Lei-Gomez Q, Mendell MJ. Meta-analyses of the associations of respiratory health effects with dampness and mold in homes. *Indoor Air Journal*, 2007, 17:284–295.
15. Mudarri D, Fisk WJ. Public health and economic impact of dampness and mold. *Indoor Air Journal*, 2007, 17:226–235.
16. *WHO guidelines for indoor air quality: dampness and mould.* Geneva, World Health Organization, 2009.
17. Zeeb H, Shannoun F, eds. *WHO handbook on indoor radon: a public health perspective.* Geneva, World Health Organization, 2009.

18. Annesi-Maesano I, Moreau D. Potential sources of indoor air pollution and asthma and allergic diseases. In: Ormandy D, ed. *Housing and health in Europe: the WHO LARES project*. Oxon/New York, Routledge, 2009.
19. *Asbestos: elimination of asbestos-related diseases*. Geneva, World Health Organization, 2010 (WHO Fact Sheet Series, No. 343). (http://www.who.int/mediacentre/factsheets/fs343/en/index.html)
20. Tossavainen A. Consensus report: asbestos, asbestosis and cancer: the Helsinki criteria for diagnosis and attribution. *Scandinavian Journal of Work, Environment & Health*, 1997, (23)4:311–316.
21. *Childhood lead poisoning*. Geneva, World Health Organization, 2010.
22. *Local housing and health action plans: a project manual*. Copenhagen, World Health Organization Regional Office for Europe, 2007.
23. Barnard LF et al. Excess winter morbidity and mortality: do housing and socio-economic status have an effect? *Reviews on Environmental Health*, 2008, 23(3):203–221.
24. *Housing, energy and thermal comfort: a review of 10 countries within the WHO European Region*. Copenhagen, World Health Organization Regional Office for Europe, 2007.
25. Wilkinson P, Armstrong B, Landon M. *Cold comfort: the social and environmental determinants of excess winter deaths in England, 1986–1996*. London, Policy Press, 2001.
26. *Euroheat: improving public health responses to extreme weather/heat-waves: summary for policy-makers*. Copenhagen, World Health Organization Regional Office for Europe, 2009.
27. Robine JM et al. Death toll exceeded 70 000 in Europe during the summer of 2003. *Comptes Rendus Biologies*, 2008, 331(2):171–178.
28. *Improving public health responses to extreme weather/heat-waves*. EuroHEAT, 2009.
29. Vandentorren S. Heat wave in France: risk factors for death of elderly people living at home. *European Journal of Public Health*, 2006, 16(6):583–591.
30. Cotton Matsui E et al. Cockroach allergen exposure and sensitization in a suburban population, *Journal of Allergy and Clinical Immunology*, 2002, 109(1), Supplement 1:89.
31. Üzel A et al. Evaluation of the relationship between cockroach sensitivity and house-dust-mite sensitivity in Turkish asthmatic patients. *Respiratory Medicine*, 2005, 99(8):1032–1037.
32. Cohn RD. Cockroach allergens: exposure risk and health effects. *Encyclopedia of Environmental Health*, 2011:732–739.
33. Sever M et al. Efficacy of extermination/intensive cleaning in reduction of cockroach allergen in inter-city North Carolina homes. *Journal of Allergy and Clinical Immunology*, 2003, 111(1), Supplement 2:205.
34. Arruda LK et al. Cockroach allergens: environmental distribution and relationship to disease. *Current Allergy and Asthma Reports*, 2001, 1(5):466–73.
35. *Innovation for health – research that makes a difference*. TDR annual report 2009. Geneva, World Health Organization, 2010.
36. Hygge S, Evans GW, Bullinger M. A prospective study of some effects of aircraft noise on cognitive performance in schoolchildren. *Psychological Science*, 2002, 13:469–74.
37. Stansfeld SA et al. Aircraft and road traffic noise and children's cognition and health: a cross-national study. *Lancet*, 2005, 365:1942–1949.
38. *Burden of disease from environmental noise. Quantification of healthy life years lost in Europe*. Copenhagen, WHO Regional Office for Europe, 2011.
39. Shield B, Dockrell JE. External and internal noise surveys of London primary schools, *Journal of the Acoustical Society of America*, 2004, 115:730–38.
40. Franssen EA et al. Aircraft noise around a large international airport and its impact on general health and medication use. *Occupational and Environmental Medicine*, 2004, 61:405–13.
41. Evans GW, Marcynyszyn LA. Environmental justice, cumulative environmental risk, and health among low- and middle-income children in upstate New York. *American Journal of Public Health*, 2004, 94:1942–44.
42. Niemann H, Maschke C. Noise effects and morbidity. In: Ormandy D, ed. *Housing and health in Europe: the WHO LARES project*. Oxon/New York, Routledge, 2009.
43. Berglund B, Lindvall T. Community noise: final report to the World Health Organization. *Archives of the Center for Sensory Research*, 1995, 2(1):1–180.
44. Batty M. The size, scale and shape of cities. *Science*, 2008, 319:769–771.
45. Saelens BE, Sallis JF, Black JB et al. Neighborhood-based differences in physical activity: an environment scale evaluation. *American Journal of Public Health*, 2003, 93:1552–1558.
46. Giles-Corti B, Donovan RJ. Relative influences of individual, social environmental and physical environmental correlates of walking. *American Journal of Public Health*, 2003, 93:1583–1589.
47. Fletcher E. Road transport, environment and social equity in Israel in the new millennium. *World Transport Policy and Practice*, 1999, 5:8–17.
48. Timperio A et al. Perceptions about the local neighborhood and walking and cycling among children. *Preventive Medicine*, 2004, 38:39–47.
49. Steinfeld E, Danford GS. Theory as a basis for research on enabling environments. In: Steinfeld E, Danford GS, eds. *Enabling environments: measuring the impact of environment on disability and rehabilitation*, New York, Kluwer Academic / Plenum Publishers, 1999:11–33.
50. Churchman A. *Differentiated perspective on urban quality of life: women, children and the elderly*. UNESCO Programme on Man and Biosphere, Proceedings of International Symposium, Rome, UNESCO, 1993.
51. Kaplan S, Kaplan R. Health, supportive environments, and the reasonable person model. *American Journal of Public Health*, 2003:93, 9:1484–1489.
52. Austin DM, Furr LA, Spine M. The effects of neighborhood conditions on perceptions of safety. *Journal of Criminal Justice*, 2002, 30:417–427.
53. Stansfeld S, Haines M, Brown B. Noise and health in the urban environment, *Reviews on Environmental Health*, 2000, 15(1–2):43–82.
54. Macintyre S, Ellaway A. Ecological approaches: rediscovering the role of the physical and social environment. In: Berkman LF and Kawachi I, eds. *Social epidemiology*. Oxford, University Press, 2000:332–348.

55. Van Poll R. *The perceived quality of the urban residential environment: a multi-attribute evaluation.* Westrom, Roermond 1997.
56. Bistrup ML. Housing and community environments – how they support health. *Briefing book for the Sundsvall Conference on Supportive Environments 1991.* Copenhagen, National Board of Health, 1991.
57. Urban features: children, slums' first casualties. In: *State of the world's cities 2006/7.* New York, United Nations Human Settlement Programme, 2007. (http://www.unhabitat.org/downloads/docs/5637_49115_SOWCR%2016.pdf)
58. *Tuberculosis.* Geneva, World Health Organization, 2010 (WHO Fact Sheet Series, No. 104). (http://www.who.int/mediacentre/factsheets/fs104/en/index.html)
59. Clark M, Riben P, Nowgesic E. The association of housing density, isolation and tuberculosis in Canadian First Nations communities. *International Journal of Epidemiology*, 2002, 31:940–945.
60. Pokhrel AK et al. Tuberculosis and indoor biomass and kerosene use in Nepal: a case control study. *Environmental Health Perspectives*, 2010, 118(4):558–564.
61. Clauson-Kaas J et al. *Crowding and health in low-income settlements.* Nairobi, United Nations Commission on Human Settlements, 1997.
62. Lawrence RJ. Housing and health: beyond disciplinary confinement. *Journal of Urban Health*, 2006, 83(3):540–549.
63. Baker M et al. Tuberculosis associated with household crowding in a developed country. *Journal of Epidemiology and Community Health*, 2008, 62:715–721.
64. Campbell-Lendrum D, Corvalan C. Climate change and developing-country cities: implications for environmental health and equity. *Journal of Urban Health: Bulletin of the New York Academy of Medicine*, 2007, 84(1):109–117.
65. Arunachalam N et al. Eco-bio-social determinants of dengue vector breeding: a multicountry study in urban and periurban Asia. *Bulletin of the World Health Organization*, 2010, 88(3):173–184.
66. *Situation update of dengue in the SEA Region, 2010.* Regional Office for South-East Asia, World Health Organization. (http://www.searo.who.int/LinkFiles/Dengue_Dengue_update_SEA_2010.pdf)
67. *Dengue guidelines for diagnosis, treatment, prevention and control.* Geneva, Special Programme for Research and Training in Tropical Diseases & World Health Organization, 2009. (http://whqlibdoc.who.int/publications/2009/9789241547871_eng.pdf)
68. Confalonieri U et al. Human health. In: Parry ML et al., eds. *Climate change 2007: impacts, adaptation and vulnerability. Contribution of Working Group III to the fourth assessment report of the Intergovernmental Panel on Climate Change, 2007.* Cambridge & New York, Cambridge University Press, 2007:392–432.
69. *Progress on sanitation and drinking water: 2010 update.* Geneva, World Health Organization & UNICEF, 2010.
70. *The health and environment linkages initiative (HELI).* Geneva, World Health Organization, 2010. (http://www.who.int/heli/en/)
71. *World health statistics 2010.* Geneva, World Health Organization, 2010.
72. Sengoelge M, Hasselberg M, Laflamme L. Child home injury mortality in Europe: a 16-country analysis. *European Journal of Public Health*, 2010.
73. *Home accident surveillance system*, 21st annual report, London, UK Department of Trade and Industry, 1999.
74. *World report on child injury prevention.* Geneva, World Health Organization, 2008.
75. Moore R. Domestic accidents. In: Ormandy D, ed. *Housing and health in Europe: the WHO LARES project.* Oxon/New York, Routledge, 2009.
76. Keall M et al. Association between the number of home injury hazards and home injury. *Accident Analysis and Prevention*, 2008, 40:887–893.
77. Keall M et al. Assessing housing quality and its impact on health, safety and sustainability. *Journal of Epidemiology and Community Health*, 2010, 64(9):765–71.
78. Fredouille J et al. Housing and mental health. In: Ormandy D, ed. *Housing and health in Europe: the WHO LARES project.* Oxon/New York, Routledge, 2009.
79. Halpern D. *Mental health and the built environment.* London, Taylor & Francis, 1995.
80. *Is housing improvement a potential health improvement strategy?* Copenhagen, Regional Office for Europe Health Evidence Network/World Health Organization, 2005.
81. *A human health perspective on climate change.* Triangle Park, Environmental Health Perspectives & National Institute of Environmental Health Sciences, 2010.

3

He Jianqing

Hunan Province, China: A double-layer aluminium roof improves the "thermal envelope" of a house in the village of Yueyang.

Evaluating health co-benefits and risks of IPCC-reviewed mitigation strategies

In this chapter, the health co-benefits of key climate change mitigation options for residential buildings are evaluated in light of the Chapter 6 of the *Contribution of Working Group III to the Fourth Assessment Report*,[1] hereafter described as the *IPCC mitigation review*. Reference here is made to key IPCC-reviewed strategies in terms of energy-efficient building design and function. Each mitigation measure is examined in terms of relevant peer-reviewed literature on health impacts, both potential co-benefits and risks.

3.1 Methods of analysis

This report presumes that improved health sector familiarity with climate change mitigation strategies is critical if maximum health co-benefits are to be gained from fine-tuning those strategies. Similarly, engineers, architects and mitigation specialists should be familiar with the health benefits and potential risks of energy efficiency and mitigation to optimize such measures. Thus citation is briefly made to key mitigation measures, followed by review of corresponding health literature, in a manner to promote such familiarity and improved intersectoral analysis.

The health-oriented literature review focused on 1) evidence of key housing-related risk factors and health reviewed above and 2) studies addressing health impacts of specific mitigation strategies that had also been considered by IPCC, e.g. health impacts of insulation and energy efficiency programmes.

3.2 Scope of mitigation issues considered

The IPCC mitigation review refers to three main principles for reducing building-related emissions: increasing the energy efficiency of buildings, reducing their energy use and shifting to renewable energy sources. In this broad context, discussion focuses on the following specific strategies reviewed:

1. Improvement of the thermal envelope of buildings (IPCC 6.4.2)
2. Heating systems, including passive solar design (IPCC 6.4.3; 6.4.6–7)
3. Cooling loads (IPCC 6.4.4)
4. Air conditioning, and heating, ventilation and air conditioning systems (HVAC) (IPCC 6.4.4–5)
5. Passive solar hot water heating and photovoltaic solar electricity (IPCC 6.4.7–8)
6. Lighting (high-efficiency) and day lighting (IPCC 6.4.9–10)
7. Household appliances and electronics (IPCC 6.6.2, 6.4.11)

Along with discussion of each of these points below, Box 3.1 summarizes the mitigation measures described as "most attractive" in different regions of the world.

3.3 Limitations of the analysis

In many cases, there were challenges "matching" relevant health and mitigation evidence. Mitigation evidence often addresses building or design strategies for climate change in different categories from the health evidence. For instance, while the mitigation literature may address single measures, e.g. thermal envelope improvements, heating, etc. health literature may relate to a mix of interventions to "improve thermal conditions," and often without explicit reference as to whether these were more or less energy-efficient. Studies undertaken in New Zealand, however, reflect pioneering work to examine health co-benefits of home improvements that improve thermal conditions and reduce energy consumption and emissions through well-defined insulation and heating interventions.

> **Box 3.1: "Most attractive" mitigation measures for buildings (IPCC 6.5.4)**
>
> 1. CO_2-saving options are largest from fuel use in developed countries and countries in transition.
> 2. Conversely, electricity savings constitute the largest potential in developing countries located in the South, where the majority of emissions in the buildings sector are associated with appliances and cooling. This distribution of potential also explains the difference in mitigation costs between developing and developed countries. The cost of shifting to more efficient appliances is quickly repaid, while building shell retrofits and fuel switching, together providing approximately half of the potential in developed countries, are more expensive.
> 3. While it is impossible to draw universal conclusions regarding individual measures and end-uses, efficient lighting technologies are among the most promising measures in buildings, in terms of both cost-effectiveness and size of potential savings in almost all countries.
> 4. In developing countries, efficient cooking stoves rank second, while the second-place measures differ in industrialized countries by climatic and geographic region.
> 5. Almost all studies examining economies in transition (typically in cooler climates) have found heating-related measures to be most cost-effective, including insulation of walls, roofs, windows and floors, as well as improved heating controls for district heat.
> 6. In developed countries, appliance-related measures are typically identified as the most cost-effective, with cooling-related equipment upgrades ranking high in warmer climates.
> 7. In terms of savings, improved insulation and district heating in colder climates and efficiency measures related to space air conditioning in warmer climates come first in almost all studies, along with cooking stoves in developing countries.
> 8. Other measures that rank high in terms of savings potential are solar water heating, efficient lighting, efficient appliances and building energy management systems.

3.4 Thermal envelope

3.4.1 Summary of IPCC-reviewed mitigation measures (IPCC 6.4.2)

"Thermal envelope" refers to a building's shell as a barrier to unwanted heat or mass transfer between a building interior and outside conditions. The IPCC review cites evidence from studies[2–4] showing that: "improvements in the thermal envelope can reduce heating requirements by a factor of two to four compared to standard practice, at a few percent of the total cost of residential buildings, and at little to no net incremental cost in commercial buildings when downsizing of heating and cooling systems is accounted for."

The IPCC review also suggests that thermal envelope effectiveness depends on (i) insulation in walls, ceiling and ground or basement floor, (ii) thermal properties of windows and doors, and (iii) the rate of exchange of inside and outside air.

Key options for improving buildings' thermal envelope are summarized as follows:

- Insulation materials – Choices need to maximize long-term thermal performance, close thermal bridges and water ingress.
- Windows – Improve thermal performance of windows via improved/multiple glazing layers, low-conductivity gases between layers, use of low-emissivity coatings on one or more glazing surfaces, use of framing materials with very low conductivity, and use of glazing that absorbs or reflects heat.
- Air leakage – Installation in walls of a continuous impermeable barrier, weather-stripping, sealing through spraying particles in ducts.

3.4.2 Health impacts of mitigation measures

Several recent studies of housing improvements have shown that retrofitting existing buildings with insulation for thermal envelope improvement can yield significant health gains in terms of reduced illness, hospitalization and days off work. In terms of economic cost-benefit analysis, these health savings could greatly exceed those from direct fuel savings and from long-term CO_2 emissions reductions (see Box 3.2).[5,6] Health could thus economically justify and "drive" mitigation policies that support greater housing energy efficiencies, at least in certain settings.

Housing and health schemes are often targeted at reducing a number of inter-related health risks that may be a factor in a range of communicable and noncommunicable diseases. For instance, both chronic and acute respiratory disease may occur as a result of exposure to indoor air pollution from space heating systems and fuels; as well as asthmas and allergies from moulds that flourish in damp and poorly heated homes; and stroke and cardiovascular disorders from exposure to temperature extremes.

Specific health outcomes may be difficult to identify, and thus are often measured in terms of terms of overall mortality or morbidity, as evidenced by doctor visits, hospitalization and days off from work or school, or by risk factors, e.g. thermal conditions, noise, etc. With these limitations in mind, the following evidence regarding health co-benefits (and when relevant risks) of certain mitigation strategies is presented.

> **Box 3.2: Driving climate policies with health benefit**
>
> One striking example of how evidence of health benefits can drive energy efficiencies is the New Zealand *Housing, insulation and health* study. Cost-benefit analysis of insulation interventions showed nearly a 2:1 benefit-cost ratio, largely due to the reduced health costs of illness and medical visits, as well as improved health equity.[5,7] These findings helped shape the policy debate in New Zealand, where the government in June 2009 announced NZ$ 323 million (US$ 221 million) over four years in funding for nationwide insulation retrofits. A cost-benefit study of the roll-out to over 76 000 houses is now under way. (See Chapter 6 case study).
>
> Empirical research into such potential health co-benefits, however, remains largely neglected. Relatively few energy improvement schemes have systematically assessed the health co-benefits of insulation improvements. Conversely, health-oriented studies may lack careful comparison of health outcomes with measurable energy efficiencies. Health studies often consider multiple housing "fixes" involving heating and insulation systems, or heating, insulation and ventilation.

3.4.2.1 Impacts of insulation improvements in terms of overall morbidity-mortality

Insulation improvements often lead to reduced exposure to temperature extremes as well as reduced dampness, which in turn have impacts on a range of health outcomes as noted above, and in particular vulnerability to stroke and cardiovascular disease, asthmas and allergies, and respiratory illness, acute and chronic. In the New Zealand housing and insulation study, a randomized community trial of insulation and other thermal envelope improvements undertaken in 1350 low-income households found significant health co-benefits in terms of reduced self-reported illness, doctor visits and hospitalizations, and days off work/school.

In the random 10 percent of households where energy use was also independently measured via electricity and gas meters, there was a net decline in energy use of 13% over the period of measurement (2001–2002), although due to the small sample size, savings were not statistically significant. A very modest, but statistically significant, 1% energy saving was found in combined assessment of savings from gas and electricity as well as self-reported use of wood and coal. This explains why the economic value of reported health gains from the efficiency measures was substantially greater than the direct energy savings, and savings in CO_2 emissions.[5]

The UK Warm Front programmes, meanwhile, evaluated only the health impacts of insulation schemes. Assessment found 0.26 months of life per person were saved in cases where insulation only was installed. Combined insulation and upgrading of the heating system results in an improvement of 0.56 months of life saved per person.[8]

Numbers and choice of window placement may also increase or reduce exposures to extreme heat or cold and dampness. For instance, one study showed that having comparatively more windows per 50 m² of living space increased the risk of health effects during heat waves in France. Location of bedrooms directly under the roof, which may be poorly insulated, was also a heat wave risk factor.[9] On the other hand, a study in Denmark found window replacement protecting against winter cold significantly reduced self-reported symptoms of joint pain, headache and neck or back pain with the intervention, although a significant number of people were lost to follow-up.[10,11, i]

3.4.2.2 Health impacts in terms of risks from inadequate ventilation

An improved thermal shell often leads to a more airtight housing space; this can also increase health impacts from indoor air pollution if rooms are not adequately ventilated.

The total ventilation rate is a measurable factor that determines occupant exposure to various contaminant sources. In some developed countries, measurable norms have been defined, often as part of commercial building standards for number of air exchanges per volume of space per hour. However, the importance of ventilation is not well understood by the public, the health community, or housing and health policy-makers.

Ventilation systems can be extremely varied, and may consist of nothing more than building leakage to provide fresh air. Even in developed countries, many single-family housing units and low-rise multi-family units do not have a planned fresh air supply system, and multi-family buildings may have unbalanced or inadequate air supply systems.

In addition, many high-rise buildings today are designed with few or no windows that can be opened manually. In older buildings, windows may be permanently sealed shut, e.g. when unit air conditioners are installed.

As noted in Chapter 2, lack of ventilation can lead to adverse health outcomes linked to increased indoor concentrations of pollutants, including radon and environmental tobacco smoke, as well as increased infection transmission.[12]

One large housing and health study showed that bronchial obstructions occurred more often in people living in housing with lower air exchange rates.[13] Another multilevel intervention study in new home construction showed that increasing the fresh air supply, coupled with heat recovery systems for exhaust air, produced statistically significant improvements in quality of life and number of days free from asthma symptoms, urgent clinical care and asthma trigger exposure.[14]

Along with exacerbating chronic exposure to indoor air pollution, inadequate ventilation of home heating systems can lead to carbon monoxide poisoning. Neurological sequelae and death occur when combustion consumes oxygen inside the house with insufficient fresh air replacement. This issue is also frequently associated with inadequate housing conditions. Since the health outcomes of CO poisoning are most severe, it has high policy relevance.[15]

The health relevance of natural ventilation is also discussed in section 3.6, with reference to cooling strategies.

> Ventilation systems may consist of nothing more than building leakage to provide fresh air. Even in developed countries, many single-family and low-rise units do not have a planned fresh air supply, and multi-family buildings may have inadequate systems.

---

[i] Reductions reported in joint pain were (OR=0.28, p<0.01), headache (OR=0.72, p<0.01) and neck or back pain (OR=0.18, p<0.01) with the intervention. However, a large number of people were lost to follow-up in this study (31% at 3 to 9 months).

3.4.2.3 Health impacts in terms of exposures to mould and other biological contaminants

The New Zealand studies note that cold and rainy-season mould growth on home interiors, which can stimulate chronic respiratory conditions, is often caused by thermal bridges (e.g. on walls). When improved insulation is combined with appropriate ventilation measures, mould and dampness in homes may be reduced, decreasing chronic respiratory conditions triggered by biological contaminants.[5]

If, however, improved insulation leads to decreased air exchange rates in airtight and highly insulated buildings, risks of various mould-related respiratory conditions can be heightened.[16] Higher humidity levels and more accumulation of microorganisms in indoor house dust may occur after the improvements. A series of German studies on housing retrofits note that indoor fungi (*A. penicillioides, A. restrictus, A. versicolor*) may develop even in very low humidity after insulation work if ventilation is not adequate.[17-20]

> Indoor fungi may develop even in very low humidity. This is another example of how adequate ventilation is a critical factor in respiratory health, overall wellness, and optimized heath benefits from thermal envelope improvements.

This is another example of how provision of adequate ventilation is a critical factor in respiratory health and overall wellness, and thus to optimized health benefits from thermal envelope improvements.

3.4.2.4 Health impacts in terms of noise exposure

Intervention studies on the health impacts of thermal envelope measures such as improved insulation and window replacement suggest that along with improving thermal conditions, these measures can also decrease noise disturbance, particularly when double-glazed or acoustic windows are used.[16]

Living in crowded neighborhoods and in substandard or poorly designed and constructed homes are commonly associated with increased noise levels inside the residence.[21] In addition, low-income homes and communities are frequently located in close proximity to airports, railroad yards, highways, industrial areas and urban commerce, where housing is may be more affordable, but noise levels comparatively high.[22]

3.4.2.5 Health impacts in terms of exposures to toxic building materials

Insulation may be categorized by its composition (material), by its form (structural or non-structural) or by its functional mode (conductive, radiative, convective). Non-structural forms include batts, blankets, loose-fill, spray foam and panels. The most commonly used insulation types are: spray polyurethane foam (SPF), insulating concrete forms, rigid panels, structural insulated panels, fibreglass batts and blankets, natural fibre, cotton batts, wool batts, loose-fill (cellulose), aerogels, straw bales, reflective insulation and radiant barriers, urea-formaldehyde foam and asbestos. Along with insulation per se, interior building materials are also part of the thermal shell and indoor environment. A comprehensive understanding of air pollutant emissions from interior building materials has developed over the years.[23] In recent guidelines on natural ventilation for infection control in health care settings, WHO recommends that, "Designers and contractors should be aware of the standards and regulations on building materials for indoor use. In particular, materials that can potentially release airborne respiratory-tract irritants should be avoided." This same principle holds for insulation materials; health issues related to a few specific fibres are briefly reviewed below.

Asbestos and other fibrous mineral silicates

Most uses of asbestos for building products in developed countries ceased by the mid-1980s. However, massive amounts of asbestos continue to be sold for use in developing country housing, and thus potentially, efforts to improve insulation from heat or extreme weather in poor countries could result in even greater asbestos use. In particular, asbestos is attractive as roof protection because it is lightweight, durable and waterproof.

As noted in Chapter 2, it is estimated that 125 million people already are exposed to asbestos in workplaces worldwide and exposures may also occur among populations and civil society groups (e.g. cleanup volunteers) who are untrained in, or unaware of, asbestos hazards.

Manufactured mineral fibres

As increased insulation use is likely to increase use of other manufactured mineral fibres (MMF), consideration should be given to any related risks. A number of major studies examine health impacts of MMF exposures during or after installation. These fibres are made from molten glass or rock (glass wool and rock wool) and can release fibrous dust on handling. Due to their prevalence, rock wool and glass wool fibres are ubiquitous in urban indoor environments of most developed cities. However, environmental levels are generally regarded as below those posing a health risk (e.g. <0.00005 fibres/cm^3).

Significantly, MMF fibres do not split into extremely fine fibres and few if any of those that become airborne will reach the deep lung. Those that do will not persist, as MMF are rapidly cleared from the lung and are soluble. The bulk of MMF products thus are regarded as posing little risk to health. They can, however, cause skin and upper respiratory tract irritation.[ii]

Synthetic vitreous fibers

Fibreglasses made of fine silica or other glass formulation fibres are the most common residential insulating material, and may be applied in rolls or "batts" of insulation pressed between the stud walls of a wood-framed house. Health and safety issues include potential cancer risk from exposure to glass fibres, formaldehyde off-gassing from the backing/resin,[24] petrochemicals in the resin and environmental health aspects of the production process.

In 1988, the World Health Organization declared fibreglass insulation potentially carcinogenic (Group 2B).[24] Subsequent epidemiologic studies by WHO's International

---

[ii] Fibre levels in buildings in which MMF has been used may be up to 0.001 f cm^3; during installation and handling of MMF materials, e.g. in loft spaces, fibre levels may rise to 0.19 f cm^3. However, levels quickly return to ambient levels when work with MMF is completed. It is estimated that over a 70-year lifetime, those exposed will inhale some 14.5 million MMF fibres from ambient or background sources. Some 15.5 million attic and eave spaces in the UK are insulated with MMF. At a fibre level of 0.0005 f cm^3, a total of about 70 million fibres would be inhaled during a 70-year lifetime with occupancy of 12 hours per day, seven days a week and 50 weeks annually. Thus some 60–70% of people in the UK will inhale about 80 million MMF fibres over a 70-year lifetime from buildings and background sources.

Agency for Research on Cancer (IARC) provided no evidence of increased risks of lung cancer or mesothelioma (cancer of the lining of body cavities) from occupational exposures during manufacture of these materials, and "inadequate evidence overall of any cancer risk." As a result, in 2001 the WHO classification of fibreglass was downgraded to Group 3 (not classifiable as to carcinogenicity in humans).[iii]

Fibreglass is also energy-intensive in manufacture. Fibres are bound into batts using adhesive binders which can contain phenol formaldehyde, a hazardous chemical known to slowly off-gas from the insulation over many years. The industry is mitigating this issue by switching to binder materials not containing phenol formaldehyde; for instance, some manufacturers offer agriculturally based binder resins made from soybean oil. Formaldehyde-free batts and batts made with varying amounts of recycled glass (some approaching 50% post-consumer recycled content) are now available in Europe.

Fibreglass (Photo: morguefile.com / kahanaboy)

3.5 Heating systems

3.5.1 Summary of IPCC-reviewed mitigation measures (IPCC 6.4.3; 6.4.6-7)

In developed countries and urban areas of some developing cities, heating is often provided by a district heating system or an on-site furnace or boiler using fossil fuels. In rural areas of developing countries, heating (when available) is generally with space heaters or stoves fueled either by biomass or coal as well as paraffin and kerosene. The IPCC mitigation review notes that significant opportunities exist in industrialized countries to combine aggressive thermal envelope measures with more efficient fossil fuel heating systems, including passive solar design, as exemplified by the European Union-certified *European Passive House Standard*.[iv]

Such measures according to IPCC, "have achieved reductions in purchased heating energy by factors of five to thirty (i.e., achieving heating levels less than 15 kWh/m^2/yr even in moderately cold climates, as compared to 220 and 250–400 kWh/m^2/yr for the average of existing buildings in Germany and Central/Eastern Europe, respectively)." Improved building energy management systems cited include co-generation of heat

---

[iii] The World Health Organization and International Agency for Research on Cancer (WHO/IARC) downgrade is consistent with the conclusion reached by the US National Academy of Sciences, which in 2000 found "no significant association between fiber exposure and lung cancer or nonmalignant respiratory disease in the MVF [man-made vitreous fiber] manufacturing environment." However, manufacturers continue to provide cancer risk warning labels on their products, apparently as indemnification against claims, and the literature should be considered carefully before determining that these risks should be disregarded. The US Occupational Safety and Health Administration (OSHA) chemical sampling page provides a summary of the risks, as does the NIOSH Pocket Guide of the US National Institute for Occupational Safety and Health.

[iv] European Passive House standard: http://eu.passivehousedesigner.de/

and electrical power, and energy storage in district heating (and cooling) systems; and combining an improved thermal envelope with a range of energy efficiencies as well as passive solar design:

- "Passive solar heating can involve extensive sun-facing glazing, various wall- or roof-mounted solar air collectors, double-facade wall construction, airflow windows, thermally massive walls behind glazing, and preheating or pre-cooling of ventilation air through buried pipes.[25-27,v]" (IPCC 6.4.3.1.)
- "District heating and cooling systems, especially when combined with some form of thermal energy storage, make it more economically and technically feasible to use renewable sources of energy for heating and cooling. Solar-assisted district heating systems with storage can be designed such that solar energy provides 30 to 95% of total annual heating and hot water requirements under German conditions (Lindenberger et al., 2000)." (6.4.6.2.)
- "By combining a high-performance thermal envelope with efficient systems and devices, 50–75% of the heating and cooling energy needs of buildings as constructed under normal practice can either be eliminated or satisfied through passive solar design." (IPCC 6.4.7)
- " 'Combisystems' are (passive) solar systems that provide both space and water heating. Depending on panel and storage tank size and on the building's thermal envelope performance, 10%-60% of combined hot water and heating demand can be met by solar thermal systems in central and northern European locations. Costs of solar heat have been 0.09–0.13 Euros/kWh for large domestic hot water systems and 0.40–0.50 Euros/kWh for combisystems with diurnal storage.[28] Worldwide, over 132 million m^2 of solar collector surface for space heating and hot water were in place by the end of 2003. China accounts for almost 40% of the total (51.4 million m^2), followed by Japan (12.7 million m^2) and Turkey (9.5 million m^2).[29]" (6.4.7.2)

This section focuses primarily on health impacts of home heating efficiencies for developed countries or grid-connected areas of emerging economies. Household use of biomass, coal, and kerosene for cooking, lighting, and other energy needs in low-income and developing country settings is addressed briefly in Sections 3.8–3.10, and in detail in the *Health in the Green Economy* report: *Co-benefits to health of climate change mitigation: Household energy sector in developing countries*.[vi]

3.5.2 Health impacts of mitigation measures

As with thermal envelope improvements, home heating and health schemes often involve multiple "fixes" to a housing structure, and are targeted at reducing a range of interrelated health risks such as: reduced chronic and acute respiratory disease from reduced exposure to indoor air pollution from space heating systems and fuels; reduced chronic respiratory disease from moulds in damp and poorly heated homes; and reduced overall mortality/morbidity from respiratory and cardiovascular conditions that may be stimulated by extremes of heat or cold.

---

[v] Technical details concerning conventional and more advanced passive solar heating techniques, real-world examples and data on energy savings are provided by Hastings (1994), Hestnes et al. (2003) and Hastings (2004).

[vi] Adair-Rohani H and Bruce N. Geneva, World Health Organization, 2011. Summary of findings available at: http://www.who.int/hia/green_economy/en/index.html

Traditionally, health-oriented studies of improved home heating examined health impacts of increased indoor warmth and thermal comfort without precise reference to gains (or losses) in energy efficiency. More recently energy efficiencies and health impacts are being looked at in tandem, but this "co-benefits" approach is relatively new.

In this context, definition of what constitutes a healthy indoor temperature vary widely. In China, the 'safe' indoor temperature range is defined in national standards as 16–24° C for winter and 22–28° C in summer. Japan's Ministry of Environment "Cool Biz" campaign encourages offices to set air-conditioners to 28° C with appropriate dress.[1] In the European region, the optimum temperatures for living spaces has been defined much more narrowly, at 20–22° C.[vii] Clearly thermal comfort is subject to influence by cultural norms about availability and access to energy, as well as differences in dress, indoor/outdoor exercise, intake of foods and drink, and other warming/cooling behaviours. Research on health co-benefits of energy efficiencies and housing mitigation strategies requires a more nuanced and up-to-date definition of healthy thermal conditions in different settings. With these limitations in mind, evidence of health co-benefits (and risks) is cited here.

Definitions of what constitutes a healthy indoor temperature vary widely and may be influenced by cultural expectations and norms, as well as by differences in dress, indoor/outdoor exercise, intake of foods and drinks and warming/cooling behaviours.

3.5.2.1 Health co-benefits from carbon-efficient home heating and fuel shifts in developed countries

New Zealand's *Housing, heating and health* study[30] was conducted in over 400 New Zealand households that had participated in an earlier study of health impacts from improved home insulation. The heating study examined health co-benefits of additional investments in more energy-efficient and non-polluting heating systems for families of children with clinical asthma symptoms. The study involved installation of energy-efficient and healthy heaters (heat pump, wood pellet burner or flued gas heater) in 400 homes of children with asthma, using either plug-in electric heaters or unflued gas space heaters – the latter often generate high levels of NOx or other air pollutants.

Along with removing certain indoor air pollutants (due to replacement of unflued gas space heaters), the measures raised average living room temperatures by an average of one degree centigrade from 16–17.1° C between 2006 and 2007 after adjustment for previous winter temperatures. The net improvement in the indoor air environment led to a statistically significant reduction in children's asthma symptoms, and reduced days off school and health care utilization.[31-33]

As with the New Zealand insulation studies, the economic benefits in terms of reduced illness that were derived from the more efficient heating systems outweighed installation costs.

While there also were net overall savings of energy and CO_2 emissions, these were relatively small, by comparison, and not statistically significant, partly due to a net increase in electricity use after the period of the heating interventions ("take-back effect"). The researchers conclude that:

[vii] For China, "Indoor Air Quality Standard (GB/T 18883-2002), as published by Ministry of Health, Ministry of Environmental Protection and Administration of Quality Supervision, Inspection and Quarantine." For the European region, as per: Ranson RP, *Guidelines for healthy housing*, Copenhagen, WHO Regional Office for Europe, 1988.

"[T]he small number of households for whom usable energy data were available also limited the statistical power of the analysis.... It also limited the extent to which we can draw conclusions about the effect of energy efficiency improvements on energy consumption. Previous research has suggested that, at indoor temperatures around 16.5°C, around 30% of potential savings following an energy efficiency improvement are likely to be "spent" as improved comfort via higher temperatures... The current results seem to suggest a higher "take-back effect," but the high degree of statistical uncertainty means that drawing strong conclusions is unwarranted."[32]

> Economic benefits in terms of reduced illness that were derived from the more efficient heating systems outweighed the installation costs. While there also were net overall savings of energy and CO_2 emissions, these were relatively small by comparison.

A number of other studies have modelled the projected health co-benefits from improved ambient or indoor air quality that could be derived from improved household heating efficiencies and/or shifts to renewable fuels in large population clusters. Two examples are cited by the IPCC:

"In China, replacement of residential coal burning by large boiler houses providing district heating is among the abatement options providing the largest net benefit per tonne of CO_2 reduction, when the health benefits from improved ambient air conditions are accounted for."[34]

A study in Greece[35] found that the residential sector's economic GHG emission abatement potential could be increased by almost 80% if the co-benefits from improved air quality are taken into account. Beyond the general synergies between improved air quality and climate change mitigation, some of the most important co-benefits in developing country households are due to reduced indoor air pollution through mitigation measures discussed in IPCC sections 6.6.2 and 6.1.1.

A series in the *The Lancet* in 2009 presented case-studies on the co-benefits to health of climate change mitigation policies. Case studies of household energy policies in both Europe and Asian countries examined:

- Impacts on health from shifts to more efficient household energy systems in the United Kingdom, including switching all indoor household fossil fuel (gas, coal, oil) combustion sources to electricity. Health impacts were significant, and greatest when combined with insulation strategies.[36]
- Impacts of low-carbon electricity grid generation in the European Union, China and India on ambient air emissions and thus health were modelled. Health impacts of a 50% emissions reduction in grid electricity generation by 2050 were assessed using an integrated modelling strategy.[37] [viii] Results indicated that a shift to low-carbon renewable energy from wind, solar, hydraulic and geothermal sources, and to nuclear energy, would reduce deaths associated with air pollution exposures (<PM2.5) in all regions. Benefit-to-cost of health gains was highest in India and China, where the health costs of pollution are high and costs of mitigation are relatively low. In the European Union, benefits also were projected at around 100 life years per million people in 2030.

[viii] This included the POLES emissions model, which identifies the distribution of production modes with desired CO_2 reductions and associated costs; the GAINS model estimating fine particulate matter (<PM2.5) concentrations); and a third model based upon WHO's Comparative Risk Assessment of mortality from outdoor air pollution, estimating the effect of <PM2.5 reductions.

3.5.2.2 Health impacts of central heating

Replacing space heating systems with central heating is not a specific focus of the IPCC mitigation review as such. However, insofar as housing and health studies often focus on this measure, it is discussed briefly here.[38,39,31]

One study of housing examined health impacts of complete central heating installation, in tandem with re-roofing, rewiring, ventilation systems, double-glazed doors, cavity wall and roof insulation. All together, the measures were reported to reduce asthma symptoms in adults, and appeared to protect against non-asthma respiratory conditions in adults and children.[40]

Another study by the UK Warm Front initiative shows improved central heating system efficiency in combination with insulation reduced the estimated prevalence of respiratory symptoms by three cases per 1000 children.[8] Increased temperatures due to installation of central heating or insulation added an extra 0.56 months to the lives of couples older than 65 years taking part in the programme: 0.33 for men and 0.22 for women. The improved thermal comfort also reduced the prevalence of depression and anxiety by 48%. The estimated prevalence of people with depression and anxiety was reduced by 150 per 1000 inhabitants.[8]

Another review of central heating interventions by the UK-based National Institute for Health and Clinical Excellence, showed a significant reduction in respiratory symptoms after the intervention as well as reductions in the number of school days lost to due to asthma-related illness (9.3 days out of 100 before intervention and 2.1 days post-intervention).[10]

> While energy efficiency may be an objective of some central heating measures, the extent to which net CO_2 emissions are reduced, as well as indoor and outdoor air pollution emissions, is not always well-defined in health-oriented review.

While energy efficiency may be an objective or a byproduct of some central heating measures, the extent to which net CO_2 emissions are reduced, as well as indoor and outdoor air pollution emissions, is not always well-defined in health-oriented reviews.

And the ease with which temperatures may be changed by central heating thermostats may also lead to reduced energy efficiency– particularly if inhabitants are unaware of, or fail to adequately control, thermostats.[41] This may pose a challenge in the optimization of both health co-benefits and energy efficiencies in central heating systems.

3.6 Cooling loads

3.6.1 Summary of IPCC-reviewed mitigation measures (IPCC 6.4.4)

Energy use for cooling can be reduced by: 1) reducing the cooling load of buildings, 2) using passive and low-energy techniques to meet some or the entire load and 3) improving the efficiency of cooling equipment and thermal distribution systems (e.g. air conditioners and vapour-compression chillers).

Principles of design cited for reducing cooling load, in most climates include:

(i) orienting a building to minimize the wall area facing east or west;
(ii) clustering buildings to provide some degree of self-shading (as in many traditional communities in hot climates);

(iii) using high-reflectivity building materials;
(iv) increasing insulation;
(v) providing fixed or adjustable shading;
(vi) using selective glazing on windows with a low solar heat gain and a high daylight transmission factor and avoiding excessive window area, particularly on east- and west-facing walls; and
(vii) utilizing thermal mass to minimize daytime interior temperature peaks.

Increasing the solar reflectivity of roofs and horizontal or near-horizontal surfaces around buildings [e.g. painting them white] and planting shade trees can also yield dramatic energy savings.[42] The benefits of trees arise both from direct shading and from cooling the ambient air. Rosenfeld et al [43] concluded that a large-scale city-wide programme of increasing roof and road albedo and planting trees in Los Angeles could yield a total savings in residential cooling energy of 50%–60%, with a 24–33% reduction in peak air conditioning loads (IPCC, 6.4.4.1).

Natural ventilation, the IPCC review notes, is one key strategy reducing the need for mechanical cooling. It functions by:

"…directly removing warm air when the incoming air is cooler than the outgoing air, reducing the perceived temperature due to the cooling effect of air motion, providing night-time cooling of exposed thermal mass and increasing the acceptable temperature through psychological adaptation when the occupants have control of operable windows.[…] Natural ventilation requires a driving force and an adequate number of openings, to produce airflow. Design features, both traditional and modern, that create thermal driving forces and/or utilize wind effects include courtyards, atria, wind towers, solar chimneys and operable windows.

"Purely passive cooling techniques require no mechanical energy input, but can often be greatly enhanced through small amounts of energy to power fans or pumps. A detailed discussion of passive and low-energy cooling techniques can be found in Harvey (2006)[44] and Levermore (2000).[45]" (IPCC 6.4.4.2)

In a housing study in Beijing, China, also cited by the IPCC review, Da Graça et al. found that thermally and wind-driven night-time ventilation could eliminate the need for air conditioning of a six-unit apartment building during most of the summer if the high risk of condensation during the day due to moist outdoor air coming into contact with the night-cooled indoor surfaces could be reduced.[46]

Evaporative cooling devices can often be used in place of air conditioners. These typically expose water to a fresh air stream, cooling the air directly or indirectly. The IPCC review notes: "By appropriately combining direct and indirect systems, evaporative cooling can provide comfortable conditions most of the time in most parts of the world." (IPCC 6.4.2.2.)

Lattice work and air flow under the roof eaves illustrates modern use of traditional natural ventilation features. (Photo: Maude Dorr)

Other passive cooling techniques cited include use of underground earth-pipe cooling, which draws outside air through a buried air duct, and dessicant dehumidification. Dessicant dehumidification, according to the review, may reduce energy required for dehumidification by 30–50% and by as much as 50–75% if solar energy is used.

> "Desiccant dehumidification and cooling involves using a material (desiccant) that removes moisture from air and that can be regenerated using heat). ... In hot and humid climates, desiccant systems can be combined with indirect evaporable cooling to provide an alternative to refrigeration-based air conditioning systems." [47]

Finally, the powerful ways in which simple measures (e.g. opening doors and windows) can increase air flow also are illustrated in a WHO-supported systematic review of the evidence on natural ventilation for infection control in health care settings (see Table 4). Findings of this review may have wider relevance in terms of residential housing, and particulation ventilation design that can help control common airborne infections in crowded housing conditions.[12]

Table 4. Estimated air changes per hour (ACH) and ventilation rate for a 7m x 6m x 3m ward

Openings	ACH	Ventilation rate (l/s)*
Open window (100%) + open door	37	1300
Open window (50%) + open door	28	975
Open window (100%) + closed door	4.2	150

* L/s (Litres per second) | *Source: (WHO, 2009)*

3.6.2 Health impacts of mitigation measures

3.6.2.1 Exposures to indoor air pollution and improved ventilation

One international review of case studies cites evidence that high-performance natural ventilation strategies can reduce respiratory illness by 9% to 20% and increase individual productivity between 0.48% and 11%, with only a minimal energy cost for increasing indoor air flow and exchange. Research on indoor air pollution-related health disorders also commonly known as "sick building syndrome" has estimated that along with better health and productivity, natural ventilation and mixed-mode conditioning can yield 25–50% energy savings.[48,49]

Based on the WHO systematic review of evidence, norms for air exchange rates in health care settings have also been developed recently by WHO.[12] This review assessed the relative risk of infection within 15 minutes of exposure to a person infected by an airborne-transmissible disease in an enclosed space at different ventilation rates, as measured by numbers of air changes per hour. Table 5, from that review, illustrates, moreover, how infection risk decreases with an increasing ventilation rate.

The basic principles of design, construction, operation and maintenance for effective natural ventilation systems, as described in the review of health care settings, have

Table 5: Infection risk in 15-minute exposure for an infector in 6m x 6.7m x 2.7m enclosed space

Quanta[ix] generation (quanta/min)	Ventilation rate (air changes per hour) (%)			
	1	6	18	30
1	0.05	0.01	0.00	0.00
7	0.30	0.06	0.02	0.01
14	0.51	0.11	0.04	0.02
20	0.64	0.16	0.06	0.04

general relevance to residential ventilation design, and may be particularly relevant to crowded housing conditions common to poor neighborhoods in developed and developing countries, and settings with high incidence of TB and/or other airborne respiratory diseases.[50]

Appropriate ventilation and air exchange rates also are important to the reduction of indoor air pollution risks that are common triggers of noncommunicable respiratory diseases, including allergies and asthmas. In a study of the health co-benefits of climate mitigation strategies for the United Kingdom of Great Britain and Northern Ireland, improved ventilation control (including mechanical ventilation and filtering of outdoor air when particle pollution concentrations outdoors were high) was found to significantly reduce indoor air pollution levels of fine particles, radon, moulds and environmental tobacco exposure.[36]

Improved ventilation control not only showed the greatest health benefit among mitigation measures considered; it also amplified health co-benefits in combined scenarios involving improved insulation, occupant behaviour, fuel switching and other measures.

Combined with natural ventilation, these measures achieved relatively larger fine particle reductions as well as significantly reduced risks of CO poisoning. If outdoor air pollution/particulate levels are higher than indoors due to industrial emissions, traffic or building-related emissions of fossil fuels or biomass, direct ventilation into the house of outdoor air can increase ingress of outdoor particle pollutants. However, mechanical ventilation can be combined with air filters to address this risk.[36,51]

3.6.2.2 Exposure to extreme heat/cold

Natural ventilation can also help avert the health impacts of air conditioning (described in more detail in 3.7.2) by reducing reliance on it as a heat-wave response measure. However, simple natural ventilation measures (e.g. window opening) will be less effective in heat waves when temperatures remain high at night. Similarly, when the temperature is above 35° C, fans alone will not necessarily prevent heat-related illness. Fans also can contribute to heat exhaustion with additional heat released indoors and forced convection during high-heat-stress conditions, when skin convection is no longer possible.[52]

[ix] Note: Quanta refers to the generation of droplet nuclei by an infected person coughing or sneezing.

In considering response to heat stress and heat waves, health literature has so far tended to consider air conditioning against simple fans or window opening/closing. Less if any consideration has been given to addressing heat stress with an energy-efficient mix of measures. As described by IPCC, these could include natural ventilation plus evaporative coolers and/or dehumidification, plus vegetation shading and other building positioning and design measures. Green roofs, for instance, can help mitigate exposure to extreme heat from heat waves.

3.6.2.3 Control of allergen-causing moulds and mites

Inadequate ventilation is also associated with greater moisture and mould risks; conversely, improved ventilation can help control moisture and mould problems. The *Lancet* case studies on health co-benefits of climate change mitigation in housing[36] noted that combined ventilation and dehumidification may help to reduce mite levels that trigger allergies. In a national survey in the United States of America, the use of a dehumidifier was an independent predictor of lower levels of mould and some asthma triggers.[53] In high-humidity settings surveyed, either a dehumidifier or an air conditioner were equally effective in reducing mite levels.

In some settings, however, evaporative coolers may promote growth of mould and mites. And one challenge in the risk assessment of fungi-related allergies is that prevalence of indoor fungi is not well mapped or defined, insofar as fungi-specific humidity indicators are not easy to test. Prevalence indicators of already-known outdoor species cannot be applied to those indoors.[54] Until this barrier is overcome, assessment of how well housing different strategies reduce fungi- and fungi-related allergies will remain difficult.

3.6.2.4 Other health impacts

Low-energy home designs and cooling systems can have other positive health impacts, supporting energy resilience during heat waves and extreme weather, which can also cause urban power loss or brownouts [55] and reducing long-term climate impacts on health from use of more carbon intensive cooling systems.

At the same time, risks exist. In regions with increasingly scarce fresh water supplies, use of some low-energy devices, such as evaporative coolers, could prove challenging.[56] In many urban areas, security concerns make it difficult to leave doors and windows open. In malaria-endemic regions, appropriate use of bednets and also house screening would be essential to any natural ventilation strategy (see Chapter 4.3.2).

3.7 Whole-building heating, ventilation and air conditioning systems (HVAC) and space/unit air conditioners

3.7.1 Summary of IPCC-reviewed mitigation measures (IPCC 6.4.4–5)

Air conditioning use is increasing rapidly around the world. In Europe, only a small fraction of residential buildings had air conditioning until recently. However, air conditioning use is rapidly increasing in response to recent heat waves. Similarly, in developing countries, IPCC notes that:

> "Until recently, the penetration of air conditioning in developing countries has been relatively low, typically only used in large office buildings, hotels and high-income homes. That is quickly changing however, with individual apartment and home air conditioning becoming more common in developing countries, reaching even greater levels in developed countries. This is evident in the production trends of typical room-to-house sized units, which increased 26% (35.8 to 45.4 million units) from 1998 to 2001 (IPCC/TEAP, 2005)." (IPCC 6.4.4.3)

morguefile.com / evildrjeff

Along with the energy used to run the air conditioning system, the halocarbon refrigerants used by air conditioners carry a high climate penalty as they are powerful greenhouse gases.

> "Air conditioners – from small room-sized units to large building chillers – generally employ halocarbon refrigerant in a vapour-compression cycle. …In some cases, the GWP-weighted lifetime emissions of the refrigerant will outweigh the CO_2 emissions associated with the electricity, highlighting the need to consider refrigerant type and handling as well as energy efficiency…" (IPCC 6.4.4.3.)

Of the key classes of halocarbon refrigerants most commonly used, HCFCs are only to be phased out in 2030–40 and HFCs remain unregulated by any global convention. Projected emissions of HFCs and HCFCs, plus ongoing emissions from CFC banks, are so large that scenarios of halocarbon emissions related to buildings in 2015 will be almost as high as in 2002[x].

In terms of mitigation potential, the IPCC review highlights several strategies: 1) reduced use of air conditioning refrigerants with the highest global warming potential; 2) more energy-efficient systems and 3) use of mixed mode natural ventilation with better temperature control.

More efficient HVAC design and management measures, the review notes, can achieve "dramatic savings in the energy use for heating, cooling and ventilation," for large buildings, such as:

---

[x] Three classes of halocarbons are mentioned in relation to air conditioning: CFCs, HCFCs and HFCs. Most consumption of ozone-destroying CFCs ended in 1996 in developed countries, while developing countries had until 2010 to phase out CFCs. HCFCs, also ozone depletors, also are being phased out of production between 2030 and 2040. Nevertheless, projected emissions of HCFCs and HFCs (and ongoing emissions from CFC banks) are sufficiently high that scenarios of halocarbon emissions related to buildings in 2015 show almost the same emissions as in 2002 (about 1.5 $GtCO_2$- eq. emissions). (IPCC 6.4.15)

Air conditioners in a building in Beijing, China.
(Photo: Matthias Braubach)

"i) using variable air volume systems to minimize simultaneous heating and cooling of air; ii) using heat exchangers to recover heat or cold from ventilation exhaust air; iii) minimizing fan and pump energy consumption by controlling rotation speed; iv) separating ventilation from heating and cooling functions…; v) separating cooling from dehumidification functions through the use of desiccant dehumidification; vi) implementing a demand-controlled ventilation system in which ventilation airflow changes with changing building occupancy, which alone can save 20 to 30% of total HVAC energy use;[57] vii) correctly sizing all components; and viii) allowing the temperature maintained by the HVAC system to vary seasonally with outdoor conditions (IPCC 6.4.5.1)

Temperature control and shifts to "mixed mode" buildings that use natural ventilation whenever possible, are the other areas of potentially large, mitigation gains:

"…a large body of evidence indicates that the temperature and humidity set-points commonly encountered in air-conditioned buildings are significantly lower than necessary,[58,59] while computer simulations by Jaboyedoff et al.[60] and by Jakob et al.[61] indicate that increasing the thermostat by 2 °C to 4° C reduces annual cooling energy use by more than a factor of three for a typical office building in Zurich, and by a factor of two to three if the thermostat setting is increased from 23° C to 27° C for night-time air conditioning of bedrooms in apartments in Hong Kong[62].

"Additional savings can be obtained in 'mixed-mode' buildings, in which natural ventilation is used whenever possible, making use of the extended range associated with operable windows, and mechanical cooling is used only when necessary during periods of very warm weather or high building occupancy." (IPCC 6.4.5.1)

3.7.2 Health impacts of mitigation measures

While the importance of using natural ventilation for control of communicable and noncommunicable disease is well documented, research into the comparative health impacts of air conditioning and/or lower-energy modes of ventilation and cooling is lacking. Similarly, there have been no studies of the comparative health impacts of conventional HVAC systems versus more energy-efficient mixed-mode HVAC systems.

A number of potential health risks, as well as several health benefits, have, however, been identified as relevant to use of air conditioning and HVAC. On the positive side, there is some evidence of that air conditioning use may improve protection from heat stress in heat waves and also be perceived as a protective measure against insect bites and vector-borne diseases. On the other hand, there is evidence that use of air conditioners, particularly large HVAC systems, may promote certain kinds of infection transmission. Such systems also create urban residential noise pollution as well as contributing to the urban heat island effect. These, in turn, have equity impacts, since residents exposed to excess noise or urban heat may be among those groups least able to afford air conditioners themselves. Some examples of available literature are reviewed below.

3.7.2.1 Immediate protection from heat waves

During the 1999 Chicago heat wave, the strongest protective factor in preventing heat-related mortality was a working air-conditioning system (OR = 0.2); however, the overall climate-appropriate design of buildings was not assessed.[63] Other evaluations have not measured the effectiveness of legislating, mandating or promoting the installation of air conditioning to prevent heat-related mortality. In relation to heat-related morbidity and mortality, there is no information on the benefits of air conditioning in relation to mortality risk in Europe, which is unsurprising given the current low coverage of this intervention.[64]

3.7.2.2 Infection transmission

Sick-building syndrome, including mucous membrane irritation, headache, fatigue and upper and lower respiratory symptoms, have been associated with HVAC systems in some studies. On average, the prevalence of such symptoms was higher in air conditioned buildings than in those naturally ventilated, independent of humidification.[65,66] Thus moisture and microbial contamination – not only of the building structure or surfaces, but also of heating, ventilation and air-conditioning systems – have adverse health effects.

Cool parts of air conditioning units can result in surface temperatures below the dew point of the air, and in damp air this may result in unwanted condensation. Microorganisms can grow in cooling-coils, drip pans, air humidifiers and cooling towers, which cause respiratory diseases or symptoms such as Legionnaires' disease and humidifier fever.[67–69] Legionellosis is a serious and sometimes fatal form of pneumonia caused by the bacterium *Legionellosis pneumophila* and other *legionella* species.

Better hygiene, commissioning, operation and maintenance of air-handling systems is particularly important in reducing such negative health impacts of HVAC systems.[64,70-74] There is evidence that temperature, humidity and air velocity of air conditioning systems are important health parameters affecting dust distribution and microbial growth.[xi, 75]

> Sick building syndrome, including mucous membrane irrigation, headache, fatigue, and respiratory symptoms have been associated with HVAC systems in some studies. On average, prevalence of such symptoms was higher in air conditioned buildings.

3.7.2.3 Outdoor air pollution exposure in "hotspots"

In homes near traffic "hotspots" or other areas of heavy outdoor air pollution concentrations, use of home air conditioning systems has been linked to reduced health impacts,[76] including asthma and allergic symptoms. A cross-sectional study of 2994 randomly selected children in areas with reportedly heavy traffic found stronger associations between asthma and rhinitis symptoms among children sleeping in non-air conditioned homes than for children sleeping in air conditioned homes.[77, xii]

3.7.2.4 Air conditioning and HVAC as informal vector control measures

In many warm regions where vector-borne diseases such as malaria and dengue are endemic, the use of air conditioning and HVAC, particularly in large-city apartment

---

xi Studies of microbial concentrations (bacteria and fungi) under different HVAC operating conditions (temperature, relative humidity and air velocity) showed that increased air velocity correlated positively with increased dust distribution and microorganism growth. Microbial growth also accelerated in the 22–32° C temperature range and with increased relative humidity (RH) between 40%–90%.

xii The International Study of Asthma and Allergies in Childhood (ISAAC) found PRs for heavy traffic density were 2.06 for wheeze (95% CI 0.97–4.38), 2.89 for asthma (1.14–7.32), 1.73 for rhinitis (1.00–2.99) and 3.39 for rhinoconjunctivitis (1.24–9.27). No associations were found for children sleeping in air-conditioned homes, suggesting that bedroom AC modifies traffic health effects among preschool children.

buildings and hotels, may be perceived as the most readily available means to prevent entry of disease-carrying mosquitoes and other vectors. This may be particularly the case at night, when temperatures typically cool down, and it might be otherwise be more healthful to use natural ventilation. While evidence of these trends has not been reviewed systematically for this paper, anecdotal observation indicates that in the absence of strong housing and health policies related to window screens and their maintenance, as well as to bednets, risk of vector-borne infections may be a powerful, but as yet unacknowledged, driver in the upsurge in air conditioning as a year-round vector control measure in urban areas and among more affluent populations.

3.7.2.5 Noise exposures

The widespread use of more energy-intensive cooling systems has secondary impacts in terms of urban noise, that stimulates a vicious cycle. As large HVAC systems and single-unit air conditioners proliferate in crowded urban areas, homes with air conditioning may opt to run their units not only for cooling purposes, but also to block street noise. Paradoxically, noise thus generated, particularly by older or improperly maintained systems, can pose a disturbance to neighbors' sleep. Noise levels generated by different air conditioning systems in dense residential areas require further study.

3.7.2.6 Equity issues

Promotion of air conditioning as a universal heat wave or indoor air pollution control measure, particularly in settings where alternative design measures can be used, ignores some key health risks discussed here, and may generate further health inequities.

This is because 1) only wealthier sectors of society can typically afford what is essentially an energy-intensive and expensive measure, and 2) increased use of air conditioning also increases heat production and thus temperatures in cities. This, in turn, increases heat-related exposures, particularly among groups with less access to air conditioning, e.g. the urban poor and elderly poor.[78]

Also, along with other measures discussed here, social networks have also been shown to help prevent heat-related deaths in heat waves, and might yield a more equitable set of health co-benefits, were such interventions to be tested systematically against the use of air conditioning.[63]

Finally, while use of conventional HVAC systems may be regarded by some as inevitable in large urban buildings, that view may soon become obsolete. Many of the measures discussed by IPCC already are being integrated into commercial as well as residential building design in developed as well as developing countries (e.g. India). These models should be closely examined in terms of how new concepts in building design and technologies can make optimized use of passive cooling and natural ventilation techniques, even in very hot climates, and in ways that may generate optimal health in a more climate-friendly built environment.

3.8 Passive and photovoltaic solar energy

3.8.1 Summary of IPCC-reviewed mitigation measures (IPCC 6.4.7.–8)

Solar photovoltaics which transform the sun's energy into electricity are highlighted in terms of their potential to meet electricity needs in well-designed buildings where passive design features have reduced overall energy demand for heating and cooling.

> "Photovoltaic panels can be supplemented by other forms of active solar energy, such as solar thermal collectors for hot water, space heating, absorption space cooling and dehumidification.

> "Building-integrated PV (BiPV) consists of PV modules that function as part of the building envelope (curtain walls, roof panels or shingles, shading devices, skylights). BiPV systems are sometimes installed in new 'showcase' buildings even before the systems are generally cost-effective. These early applications will increase the rate at which the cost of BiPVs comes down and the technical performance improves..."[79]

> "Gutschner et al. (2001)[80] have estimated the potential for power production from BiPV in IEA member countries. Estimates of the percentage of present total national electricity demand that could be provided by BiPV range from about 15% (Japan) to almost 60% (USA)." (IPCC 6.4.7.1)

As noted in Section 3.5, passive solar hot water heaters (usually consisting of plates or tubing through which hot water flows) are becoming increasingly popular, including in emerging economies such as Turkey and China. Solar hot water heating can provide an estimated 50-90% of annual hot water needs, depending on climate (IPCC 6.4.8).

3.8.2 Health impacts of mitigation measures

3.8.2.1 Reduced energy poverty and related health risks

The IPCC review focuses much discussion on cutting-edge PV technologies being tested in large buildings in high-income or industrialized settings. In the context of health co-benefits, however, the adoption and use of simple solar technologies in poor homes may be even more relevant.

Small off-grid photovoltaic and passive solar systems are becoming more widely available in Asia, Africa and Latin America – particularly for domestic lighting and domestic hot water heating (this latter trend is noted by IPCC). As household- and community-based energy options develop, these may "leap-frog" over weak or non-existent grid-connected systems in some regions, much as mobile phone technologies supplanted fixed-line phone systems a decade ago.

In India, simple but high-efficiency solar photovoltaic lanterns are being used by tens of thousands of poor households to generate lighting and other household needs; these are often recharged at a local "PV recharge" station that operates as a micro-business. (See Case Study, *Lighting a Billion Lives,* Chapter 6.2).

An Indian woman takes her solar-powered lantern to a local PV station where she can recharge the battery for a small sum. (Photo: TERI/India)

In South Africa, a housing upgrade project in the Kuyasa neighbourhood of Cape Town, supported by finance through the United Nations Framework Convention on Climate Change (UNFCC), Clean Development Mechanism (CDM) incorporates passive solar technology for hot water heating, into more energy-efficient home retrofits. (See Case Study, *Low-cost urban housing*, Chapter 6.3).

The health co-benefits of passive and active solar technologies for poor households lacking ready access to electricity and hot water are increasingly appreciated at the grassroots. But these have not been systematically assessed; this remains a serious gap in the housing and health literature. Anecdotal evidence, however, indicates benefits in terms of: improved sanitation through more access to hot water for kitchen and personal hygiene; reduced risks of burns and indoor air pollution related to the use of kerosene and paraffin lights; increased sense of security and ability to study, work and move around the house at night. Just a few indicative examples are noted here:

In terms of hot water access, the use of hot dish water at sufficiently high temperatures (<40° C) has been cited in some studies as a factor reducing transmission of common bacteria (*E. coli*, *salmonella*, and *campylobacter*) via dishes and utensils.[81] A study on domestic kitchen hygiene in Peru notes that even when oil- or electric-powered water heating is available, fuel poverty may be a barrier to the use of hot water for dish-washing, stating: "none of our respondents used hot water to wash their dishes, probably because of fuel costs."[82] Also, storage of water at or above 50° C can help prevent microbial build up in household water systems, including *legionella* bacteria.[69, xiii]

In terms of lighting, there is also tentative evidence that in comparison to kerosene, electricity use from alternative sources may reduce the risk of respiratory disease. One recent case-controlled study in Nepal women using kerosene lighting and had an increased risk of TB – possibly as a result of time spent in close proximity to the flame.[83]

This hospital-based case-controlled study among 375 women in Nepal who used biomass for heating and cooking, and kerosene lamps for light, found that women who used kerosene lamps for lighting had a higher overall risk of developing TB than women who used biomass stoves – but had other means of lighting their homes (e.g. electricity), the study notes. "If kerosene lamp use is a risk factor for TB, it would provide strong justification for promoting clean lighting sources, such as solar lamps."[83]

In developed countries, where shifts to renewable energy may be undertaken at power-grid level, consequent increases in energy costs may pose a disproportionate burden on low-income families. Off-grid or independent home installation and use of solar photovoltaics also, too often, remains a consumer choice available primarily to stronger socio-economic sectors. Affordability is thus an essential consideration in strategies, especially for those with low incomes.

[xiii] In principle water must be stored either below 25° C or above 50° C to prevent microbial build-up. At the same time, at tap point, water above 50° C may pose a risk of scalding, particularly of children and other vulnerable groups.

3.8.2.2 Potential occupational and environmental hazards related to PV panel production and waste disposal

Increased production of certain kinds of photovoltaic solar cells can lead to increased occupational exposures or exposures to waste products, e.g. from discarded panels. For example, cadmium-tellurium (CdTe) compounds are found in some photovoltaic systems, increasing the potential for cadmium emissions from mining and refining and from manufacture, utilization and disposal of photovoltaic modules.

Cadmium and cadmium compounds like CdTe are classified as known human carcinogens. Acute exposure to CdTe can result in respiratory irritation and toxicity. Other hazardous materials that may be present in various solar products' manufacturing include arsenic compounds, carbon tetrachloride, hydrogen fluoride, hydrogen sulphide, and lead and selenium compounds. Many of these have been linked with multiple health effects, including cancer.[84]

New solar PV production processes are being developed that make use of more innocuous compounds, such as embedded copper and silicon. Passive solar hot water heating systems typically use innocuous substances.

Workers prepare to affix a solar panel to the roof of a rural home in Sri Lanka (Photo: © World Bank)

Nanotechnology offers promise of making solar technology more efficient and affordable by improving the efficiency of batteries and other components. Yet little, so far, is known about nanotechnology's impact on health,[85] so research in this area needs to be a priority.

Finally, while there is certainly concern regarding unexplored occupational and waste-related health risks from solar and renewable technologies, the evidence is limited.[86] As evidence develops, it also must be weighed against the large occupational and environmental health risks associated with the extraction, transport and use of fossil fuels for domestic energy use,[36] as well as current knowledge about environmental health and safety risks associated with nuclear energy generation,[87] and lessons learnt in the wake of the April, 2011 Fukushima, Japan, nuclear disaster.[xiv]

3.9 Lighting and day lighting

3.9.1 Summary of IPCC-reviewed mitigation measures (IPCC 6.4.9–10)

Overall, efficient lighting technologies are cited by the IPCC mitigation review as among the most promising housing measures worldwide, in terms of both cost-effectiveness and size of potential savings, as estimated by the International Energy Agency.[88] States the IPCC review:

> "Lighting energy can be reduced by 75% to 90% compared to conventional practice through (i) use of day lighting with occupancy and daylight sensors to dim and switch off electric lighting; (ii) use of the most efficient lighting devices available; and (iii) use of such measures as ambient/task lightning." (IPCC 6.4.9)

[xiv] Note: On 19 April 2011, United Nations Secretary-General Ban Ki-moon called for a worldwide review of nuclear energy safety standards, including renewed cost-benefit analysis of nuclear energy. Office of the UN Secretary General/SG/T/2873. 25 May 2011. http://www.un.org/News/Press/docs/2011/sgt2783.doc.htm

> "…Day lighting systems involve the use of natural lighting for the perimeter areas of the building. (…) Opportunities for day lighting are strongly influenced by architectural decisions early in the design process, such as building form; the provision of inner atria, skylights and clerestories; and the size, shape and position of windows." (IPCC 6.4.10)

The review also briefly notes the equity benefits that can be obtained from mitigation measures that replace fossil-fuel based lamps with more efficient PV solar-powered or compact flourescent (CFL) electric lights:

> "About one third of the world's population depends on fuel-based lighting (such as kerosene, paraffin or diesel), contributing to the major health burden from indoor air pollution in developing countries. While these devices provide only 1% of global lighting, they are responsible for 20% of the lighting-related CO_2 emissions and consume 3% of the world's oil supply. (IPCC 6.4.9.1.)

CFL lightbulbs save energy but require careful disposal of broken bulbs, which contain mercury.

3.9.2 Health impacts of mitigation measures

Generally speaking, provision of adequate light, both natural and artificial, is a determinant of health. However, trade-offs are involved. Windows are an important source of ventilation, and adaptations in their size and positioning to reduce direct sun exposure can be an effective mitigation measure as well as reducing heat stress in heat waves. The health literature also shows that dwellings and their residents require sufficient natural light exposure as a factor in mental health, biophysical performance and injury prevention. These issues are briefly considered below.

3.9.2.1 Basic metabolic functions and health

Exposure to the outdoors and to natural light sources is essential to health and Vitamin D production. In addition, the absence of a regular natural light/dark cycle can influence body rhythms such as sleep patterns, ovulation and hormone secretion; these affect performance, alertness and mood. Lack of natural light exposure therefore may be related to symptoms of stress and depression, particularly in parts of the world where people spend most of their time indoors. There is no definitive information on the amount and duration of exposure to light that may be necessary to prevent such effects, or on the number of people who may be adversely affected. Windows may have additional psychological benefits unrelated to the provision of daylight, perhaps being associated with contact with the outside world and variety of visual stimulation, particularly for the elderly.

In temperate climates where much time is spent indoors, it has been recommended that home interiors be designed so that they receive at least 25% of probable sunlight hours (the long-term average of the total hours during the year in which direct sunlight reaches unobstructed ground). British Standards[89] recommend minimum glazed areas for a satisfactory view when windows are restricted to one wall. These areas vary from 20% to 35%, depending on the depth of the room. The same standard also defines features comprising a "good" window view in the following terms:

> "It should give information about three 'layers' (the sky, the horizon and the ground); it should be complex rather than simple; it should be changing or varied and should preferably have natural elements such as water or trees." [90]

3.9.2.2 Injury prevention and environmental safety

Studies show that inadequate lighting and day lighting can lead to injuries from falls in older adults. Inadequate exterior lighting is linked to increased risk or perceived risk of crime and attack.[91]

At the same time, as use of CFL lights becomes very widespread, care must be taken in their handling as they contain mercury that quickly gasifies and is thus released into the home environment following breakage. Because the use of CFLs will reduce demand for electricity (often generated through coal burning that also releases mercury), there a net reduction in overall mercury releases can still be achieved with CFL light use. At the same time, use of CFL lights has raised new public health concerns over direct mercury exposure, particularly of children, in the home environment. Opening the window and leaving the room for 15 mintues following a lightbulb breakage, as well as sealing the broken bulb in a plastic bag, are among the measures recommended by national agencies like the United States Environment Protection Agency.[xv]

3.9.2.3 Depression

Research shows linkages between depression and inadequate daylight exposure in housing. At the same time, the causal pathway of impact on depression is complex and may be mitigated by other factors, such as satisfaction with the dwelling, predictability of daily routines and feelings of safety, comfort and control. This complexity was noted in a study on light exposure among depressed populations: "Given the complex causal web, we would expect interaction or mediation between the variables."[92]

3.10 Household appliances and electronics

3.10.1 Summary of IPCC-reviewed mitigation measures (IPCC 6.6.2; 6.4.11)

The focus of IPCC review is greater energy efficiencies in household appliances and electronics, which represent: "more than 40% of total residential primary energy demand in 11 large OECD nations. [93-96] ...Appliances in developing countries constitute a smaller fraction of home energy demand. However, rapid uptake in emerging economies such as China, reflects their growing importance in the developing world as economies grow.[97]" (IPCC 6.4.11)

For developing countries, biomass and coal cookstoves are a key home appliance, imposing a heavy climate penalty: "If products of incomplete combustion (PICs) other than methane and N2O are considered…biomass stove-fuel combinations exhibit GHG emissions three to ten times higher than fossil-fuel alternatives."(IPCC 6.4.3) Advanced biogas or biomass cookstoves offer the largest climate mitigation potential as well as health gain from "clean domestic energy services, including safe cooking." (IPCC 6.6.2)

[xv] *Children's exposure to mercury compounds.* Geneva, World Health Organization, 2010.

3.10.2 Health impacts of mitigation measures

3.10.2.1 Indoor air pollution exposures

Cleaner biomass/biogas stoves and fuels reduce exposures to the most health-damaging emissions of smoke (e.g. particulate matter) by as much as 90%. The important health co-benefits of climate change mitigation in this sector is the focus of another report of this *Health in the Green Economy* series,[i] and are briefly summarized here. (Box 3.3).

> **Box 3.3: Health co-benefits of low-emission stoves and fuels**
>
> New technologies for more efficient household fuel use in developing regions hold some of the greatest potential co-benefits for both health and climate. New stove technologies and cleaner liquid and gaseous fuels that substantially reduce climate change emissions (e.g. CO_2, methane and black carbon particles) also reduce exposures to the most health-damaging air pollutants (e.g. particulate matter) by as much as 90%. These interventions offer co-benefits for health, gender equity and sustainable development for billions of people.
>
> The increasing synergy between cost-effective stove and fuel technologies and health gain potential is the focus of a complementary *Health in the Green Economy* report, *Co-benefits to health of climate change mitigation: the household energy sector in developing countries.*[i] This report evaluates mitigation options for household energy assessed in terms of health benefits and risks using two approaches. The first approach draws on an extensive review of laboratory and field testing in a schematic summary of overall health and mitigation benefits for a range of available fuel and technology combinations, including consideration of costs and any potential limitations or tradeoffs.
>
> This is followed by scenario-based estimates of health gains from the adoption of cleaner stoves in Latin America and sub-Saharan Africa over the next decade at a pace consistent with UN targets addressing energy poverty.
>
> Close to one million deaths could be avoided, including from childhood pneumonia, and over a longer time frame, from COPD and ischaemic heart disease in adult populations, the scenario assessments show. The study also illustrates as the mitigation potential resulting from adoption of the most promising low-emission household energy technologies for populations in sub-Saharan Africa and continental Latin American countries.
>
> This review highlights the climate-changing role of long- and short-lived pollutants that result from inefficient energy use in developing country households. The serious health impacts that arise from emissions of shorter-lived pollutants, estimated at almost two million premature deaths for the year 2004, underlines the global opportunity to achieve large health gains through mitigation measures.
>
> ---
> [i] Adair-Rohani H and Bruce N. *Health in the Green Economy, Co-benefits to health of climate change mitigation: the household energy sector in developing countries.* Geneva, World Health Organization, 2011. http://www.who.int/hia/green_economy/en/index.html

3.10.2.2 Impacts of improved heating appliances on home injuries

In developed as well as developing countries, energy efficiency measures that improve the quality and standard of heating and cooking appliances reduce home injuries, and may contribute to better indoor environment and temperature control.[98]

3.10.2.3 Improved access in developing countries to low-energy appliances, particularly refrigeration

In developed economies more efficient appliances may reduce household energy costs. For households in parts of the developing world where there is less electricity access, the emergence of low-energy appliances can put previously unattainable devices within reach. This, in turn, may have impacts on aspects of health and health equity. Of particular relevance are DC (direct current) appliances that may be powered directly by PV-solar electricity, avoiding energy losses through conversion to alternating (AC) current. Examples include computers, phones, and even refrigerators. Notably, DC-solar powered refrigerators are now being introduced into the health sector. Such devices, if adapted to domestic uses could have health impacts, e.g. for food safety, and deserve further exploration by the health community.

Table 6 (Table 6.1 of the IPCC mitigation review) maps the broad applicability of energy-efficient housing technologies in different regions. One important future challenge for the health researchers is a parallel mapping of health co-benefits by region.

A Nepalese woman cooks on a clean biogas stove, part of project supported by the local NGO "World in Justice."
(Photo: Heather Adair Rohani)

Table 6 (IPCC Table 6.1). Applicability of energy efficiency technologies in different regions

Energy efficiency or emission reduction technology	Developing countries						OECD						Economies in transition, Continental			Reference
	Cold climate			Warm climate			Cold climate			Warm climate						
	Tech stage	Cost/ effectiveness	Appropriate-ness	Tech stage	Cost/ effectiveness	Appropriate-ness	Tech stage	Cost/ effectiveness	Appropriate-ness	Tech stage	Cost/ effectiveness	Appropriate-ness	Tech stage	Cost/ effectiveness	Appropriate-ness	
Structural insulation panels	●	●	●	●	●	○	●	●	●	●	●	●	●	●	●	6.4.2.1
Multiple glazing layers	●	●	●	●	●	●[1] ○[2]	?	●	●	●[11]	●	●	●	●	●	6.4.2.2
Passive solar heating	●	●	●	●	●	○	●	●	●				●	●	●	6.4.3.1
Heat pumps	●[3]	○	●	●[4] ●	●[5] ○[6]	●[7] ○[8]	●[9]	●	●	●[10] ?	●[12] ●[13]	●[14] ●[15]	●[16]	●	●	6.4.3.3 / 6.4.8
Biomass-derived liquid fuel stove	●	●	●	?	○	●	?	●	○	?	●	○	?	●	●	6.4.3.2
High-reflectivity building materials	○	○	○	●	●	●	●	●	●	?	●	●	●	●	●	6.4.4.1
Thermal mass to minimize daytime interior temperature peaks	?	●	●[17] ○[18]	?	●	●	?	●	●	?	●	●[19] ●[20]	●	●	●	6.4.2
Direct evaporative cooler	?	●	●[21] ○[22]	●	●	●	?	○	○	?	●	●[23] ●[24]	?	○	○	6.4.4.2
Solar thermal water heater	●	●	●	●	○	●	?	●	●	●	●	●	●	●	●	6.4.8
Cogeneration	●	●	●	●	○	○	?	●	●	●	●	●	●	●	●	6.4.6
District heating and cooling system	●	●	●	●	○	○	?	●	●	●	●	●	●	●	●	6.4.6
PV	●	●	●	●	○	○	?	●	●	●	●	●	●	●	●	6.4.7.1
Air to air heat exchanger	●	●	●	●	●	●	●	●	●	●	●	●	●	●	●	6.4.5.1
High-efficiency lighting (FL)	?	●	●	?	●	●	μ	○	●	μ	○	●	●	●	●	6.4.10

72 Health co-benefits of climate change mitigation – Housing sector

Energy efficiency or emission reduction technology	Developing countries						OECD						Economies in transition, Continental			Reference
	Cold climate			Warm climate			Cold climate			Warm climate						
	Tech stage	Cost/effectiveness	Appropriate-ness	Tech stage	Cost/effectiveness	Appropriate-ness	Tech stage	Cost/effectiveness	Appropriate-ness	Tech stage	Cost/effectiveness	Appropriate-ness	Tech stage	Cost/effectiveness	Appropriate-ness	
High-efficiency lighting (LED)	~	R	E	~	R	E	R	R	E	R	R	E	R	D	E	6.4.10
HC-based domestic refrigerator	D	D	E	D	D	E	E	E	E	E	R[25,26]	E	E	D	E	6.4.11
HC or CO_2 air conditioners	R	~	μ	R	~	μ	R	R	D	R	R	E[μ27, μ28]	R	~	D	6.4.4.3
Advanced supermarket technologies	~	E	E	~	E	E	~	E	E	D	E	E	D	E	E	6.4.12
Variable speed drives for pumps and fans	~	D	D	~	D	D	~	D	D	~	D	D	~	D	D	6.4.4.2
Advanced control system based on BEMS	E	D	D	E	D	D	E	D	D	E	D	D	E	D	D	6.4.4.6

Selected are illustrative technologies, with an emphasis on advanced systems, the rating of which is different between countries.

Visual representation	Tech stage	Cost/effectiveness	Appropriateness
(light blue)	Research phase (including laboratory and development)	[R] Expensive/Not effective [$$/-]	Not appropriate [-]
(teal)	Demonstration phase [D]	Expensive/Effective [$$/+]	Appropriate [+]
(dark purple)	Economically feasible under specific conditions [E]	Cheap/Effective [$/+]	Highly appropriate [++]
	Mature Market (widespread commercially available without specific governmental support) [M]	'~' Not available	'~' Not available
~			
μ	No Mature Market (not necessarily available/not necessarily mature market)		

Notes:

1. For heat block type;
2. For Low-E;
3. Limited to ground heat source etc.;
4. For air conditioning;
5. For hot water;
6. For cooling;
7. For hot water;
8. For cooling;
9. Limited to ground heat source, etc.;
10. For cooling;
11. For hot water;
12. For hot water;
13. For cooling;
14. For hot water;
15. For cooling;
16. Limited to ground heat source, etc.;
17. In high-humidity region;
18. In arid region;
19. In high humidity region;
20. In arid region;
21. In high humidity region;
22. In arid region;
23. In high-humidity region;
24. In arid region;
25. United States;
26. South European Union;
27. United States;
28. South European Union.

Source: Climate Change 2007: Mitigation of Climate Change. Working Group III Contribution to the Fourth Assessment Report of the Intergovernmental Panel on Climate Change. Cambridge University Press, Cambridge and New York, 2007. (Table 6.1)

References

1. Levine M et al. Residential and commercial buildings. In: Metz B et al., eds. *Climate change 2007: mitigation of climate change. Contribution of working group III to the fourth assessment report of the Intergovernmental Panel on Climate Change, 2007.* Cambridge & New York, Cambridge University Press, 2007:387–446.

2. Demirbilek FN et al. Energy conscious dwelling design for Ankara. *Building and Environment*, 2000, 35:33–40.

3. Hamada Y et al. Development of a database of low-energy homes around the world and analysis of their trends. *Renewable Energy*, 2003, 28:321–328.

4. Hastings SR. Breaking the "heating barrier": learning from the first houses without conventional heating. *Energy and Buildings*, 2004, 36:373–380.

5. Chapman R et al. Retrofitting houses with insulation: a cost-benefit analysis of a randomized community trial. *Journal of Epidemiology and Community Health*, 2009, 63(4):271–277.

6. Jakob N, Nutter S. Marginal costs, cost dynamics and co-benefits of energy efficiency investments in the residential buildings sector. In: *ECEEE 2003 summer study: time to turn down energy demand*. Stockholm, European Council for an Energy Efficient Economy, 2003:829–841.

7. Howden-Chapman P et al. Effect of insulating existing houses on health inequality: cluster randomized study in the community. *British Medical Journal*, 2007, 334:460.

8. Green G, Gilbertson J. *Warm front, better health: health impact evaluation of the warm front scheme*. Sheffield, Sheffield Hallam University, 2008.

9. Vandentorren S. Heat wave in France: risk factors for death of elderly people living at home. *European Journal of Public Health*, 2006, 16(6):583–591.

10. Taske N et al. *Housing and public health: a review of reviews of interventions for improving health*. Evidence briefing. London, National Institute for Health and Clinical Excellence, 2005.

11. Iverson M, Bach E, Lunqvist G. Health and comfort changes among tenants after retrofitting of their housing. *Environment International*, 1986,12:161–6.

12. Atkinson J et al. *Natural ventilation for infection control in health-care settings*. Geneva, World Health Organization, 2009.

13. Oie L et al. Ventilation in homes and bronchial obstruction in young children. *Epidemiology*, 1999, 10(3):294–299.

14. Takaro T et al. Clinical response in asthma from improved housing design and construction. Paper presented at American Thoracic Society Annual Meeting, Toronto. *American Journal of Respiratory and Critical Care Medicine*, 2008, 177:A.

15. Summary of expert group agreements. In: *WHO second technical meeting on quantifying disease from inadequate housing*. Geneva, World Health Organization, 2006:5. (http://www.euro.who.int/__data/assets/pdf_file/0006/98673/EBD_report.pdf)

16. *Preliminary results of the WHO Frankfurt housing intervention project*. Copenhagen, World Health Organization Regional Office for Europe, 2008.

17. Trautmann C et al. Background concentrations of molds in house dust: Determination of mold concentrations in dwellings without known mold infestations in three parts of Germany. *Bundesgesundheitsblatt Gesundheitsforschung Gesundheitsschutz (Federal Health Gazette for Health Research and Health Protection)*, 2005, 48:29–35.

18. Trautmann C et al. Background concentrations of molds in air. Determination of mold concentrations in dwellings without known mold infestations in three parts of Germany. *Bundesgesundheitsblatt Gesundheitsforschung Gesundheitsschutz (Federal Health Gazette for Health Research and Health Protection)*, 2005, 48:12–20.

19. Fischer G et al. Indoor fungi and fungal allergens – possibilities and limitations of allergy diagnosis and exposure assessment. In: De Oliveira Fernandes E et al., eds. *Proceedings of Healthy Buildings*. Portugal, 2006, 1:175–180.

20. Pitten FA et al. Schimmelpilzbelastungen in Innenräumen – Befunderhebung, gesundheitliche Bewertung und Maßnahmen [Indoor fungi – evaluation, health risk assessment and interventions]. *Bundesgesundheitsblatt Gesundheitsforschung Gesundheitsschutz [Federal Health Gazette for Health Research and Health Protection]*, 2007, 50:1308–1323.

21. DiGuiseppi C et al. Housing interventions and control of injury-related structural deficiencies: a review of the evidence. *Journal of Public Health Management & Practice*, 2010, 16(5):S34–43.

22. *Healthy transport in developing cities, health and environment linkages initiative (HELI)*. Geneva, World Health Organization & United Nations Environment Programme, 2009. (http://www.who.int/heli/risks/urban/urbanenv/en/index.html)

23. Levin H. Building materials and indoor air quality. In: Hodgson M, Cone J, eds. *State of the Art Reviews in Occupational Medicine*, Vol. 4, 4, 1989.

24. *Formaldehyde*. IARC monographs on the evaluation of carcinogenic risks to humans, volume 88. Lyon, International Agency for Research on Cancer / World Health Organization. (http://monographs.iarc.fr/ENG/Monographs/vol88/mono88-6.pdf)

25. Hastings SR. *Passive solar commercial and institutional buildings: a sourcebook of examples and design insights*. Chichester, John Wiley, 1994.

26. Hastings SR. Breaking the 'heating barrier'. Learning from the first houses without conventional heating. In: *Energy and Buildings*, 2004, 36: 373–380.

27. Hestnes AGR, Hastings SR, Saxhof B. *Solar energy houses: strategies, technologies, examples*. Second Edition. London, James & James, 2003.

28. Peuser FA, Remmers KH, Schnauss M. *Solar thermal systems: successful planning and construction*. James & James & Solarprazis, Berlin, 2002.

29. Weiss W, Bergmann I, Faninger G. *Solar heating worldwide, markets and contribution to the energy supply 2003*. IEA Solar Heating and Cooling Programme, Graz, Austria, 2005.

30. Howden-Chapman P et al. Effects of improved home heating on asthma in community dwelling children: randomised controlled trial. *British Medical Journal*, 2008, 337:852–855.

31. Free S et al. Does more effective home heating reduce school absences for children with asthma? *Journal of Epidemiology and Community Health*, 2010, 64:379–386.

32. Preval N et al. Evaluating energy, health and carbon co-benefits from improved domestic space heating: a randomised community trial. *Energy Policy*, 2010, 38:3965–3972.

33. Gillespie-Bennett J et al. Sources of nitrogen dioxide (NO2) in New Zealand homes: findings from a community randomized controlled trial of heater substitutions. *Indoor Air*, 2008, 18(6):521–528.

34. Mestl SH et al. Cleaner production as climate investment: integrated assessment in Taiyuan City, China. *Journal of Cleaner Production*, 2005, 13:57–70.

35. Mirasgedis S et al. CO_2 emission reduction policies in the Greek residential sector: a methodological framework for their economic evaluation. *Energy Conversion and Management*, 2004, 45:537–557.

36. Wilkinson P et al. Public health benefits of strategies to reduce greenhouse-gas emissions: household energy. *Lancet*, 2009, 374(9705):1917–1929.

37. Haines A, et al. Public health benefits of strategies to reduce greenhouse-gas emissions: overview and implications for policy makers. *Lancet*, 2009, 374:2104–14.

38. Hopton JL, Hunt SM. Housing conditions and mental health in a disadvantaged area in Scotland. *Journal of Epidemiology & Community Health*, 1996, 50(1):56–61.

39. Somerville M et al. Housing and health: does installing heating in their homes improve the health of children with asthma? *Public Health*, 2000, 114:434–9.

40. Barton et al. The Watcombe housing study. *Journal of Epidemiology and Community Health*, 2007, 61:771–777.

41. Critchley R et al. Living in cold homes after heating improvements: evidence from England's Home Energy Efficiency Scheme. *Applied Energy*, 2007, 84:147–158.

42. *White roofs may successfully cool cities.* National Science Foundation, 2010 (http://www.nsf.gov/news/news_summ.jsp?org=NSF&cntn_id=116283).

43. Rosenfeld AH et al. Cool communities: strategies for heat island mitigation and smog reduction. *Energy and Buildings*, 1998, 28:51–62.

44. Harvey LDD. *A handbook on low-energy buildings and district energy systems: fundamentals, techniques, and examples.* London, James and James, 2006.

45. Levermore GJ. *Building energy management systems; application to low-energy HVAC and natural ventilation control*, 2nd ed. London, E&FN Spon, Taylor & Francis Group, 2000.

46. Da Graça GC et al. Simulation of wind-driven ventilative cooling systems for an apartment building in Beijing and Shanghai. *Energy and Buildings*, 2002, 34:1–11.

47. Belding WA and Delmas MPF. Novel desiccant cooling system using indirect evaporative cooler. *ASHRAE Transactions*, 1997, 103(1):841–847.

48. Fisk WJ, Rosenfeld AH. Estimates of improved productivity and health from better indoor environments. *Indoor Air*, 1997, 7:158–172.

49. Kroeling P. Health and well-being disorders in air-conditioned buildings; comparative investigations of the "building illness" syndrome. *Energy Buildings*, 1988, 11:277–282.

50. Baker M et al. Tuberculosis associated with household crowding in a developed country, *Journal of Epidemiology and Community Health*, 2008, 62:715–21.

51. Krieger J et al. Housing interventions and control of asthma-related indoor biologic agents: a review of the evidence. *Journal of Public Health Management & Practice*, 2010, 16(5S):S11–S20.

52. Menne B et al. Indoor heat protection measures and human health. In: Matthies F, Menne B, eds. *Preparedness and response to heat-waves in Europe, from evidence to action: Public health response to extreme weather events.* Copenhagen, WHO Regional Office for Europe, in press.

53. Salo PM et al. Dustborne *alternaria alternata* antigens in US homes: results from the national survey of lead and allergens in housing. *Journal of Allergy and Clinical Immunology*, 2005, 116(3):623–629.

54. Hollbach N et al. Protein spectra of conidia and mycelium of *Penicillium chrysogenum* and *Aspergillus fumigatus* wild strains and consequences for allergy diagnostics of indoor fungi. In: De Oliveira Fernandes E et al., eds., *Proceedings of Healthy Buildings, Portugal*, 2006, 2:419–424.

55. Roaf S, Crichton D, F Nicol. *Adapting buildings and cities for climate change: a 21st century survival guide*, 2nd ed. Amsterdam, Elsevier, 2009.

56. Matthies F, Menne B, eds. *Preparedness and response to heat-waves in Europe, from evidence to action. Public health response to extreme weather events.* Copenhagen, World Health Organization Regional Office for Europe, 2007.

57. Brandemuehl MJ, Braun JE. The impact of demand-controlled and economizer ventilation strategies on energy use in buildings. *ASHRAE Transactions*, 1999, 105(2):39–50.

58. De Dear RJ, Brager GS. Developing an adaptive model of thermal comfort and preference. *ASHRAE Transactions*, 1998, 104(1A):145–167.

59. Fountain ME et al. An investigation of thermal comfort at high humidities. *ASHRAE Transactions*, 1999, 105: 94–103.

60. Jaboyedoff P et al. Energy in air-handling units: results of the AIRLESS European project. *Energy and Buildings*, 2004, 36 : 391–399.

61. Jakob M et al. *Grenzkosten bei forcierten Energie-Effizienzmassnahmen und optimierter Gebäudetechnik bei Wirtschaftsbauten* [Marginal costs of energy efficiency measures and improved technology for commercial buildings]. Zürich/Bern, CEPE ETH Zurich, Amstein+Walthert, HTA Luzern / Swiss Federal Office of Energy, 2006.

62. Lin Z, Deng S. A study on the characteristics of nighttime bedroom cooling load in tropics and subtropics. *Building and Environment*, 2004, 39:1101–1114.

63. Klinenberg E. *Heat wave: a social autopsy of disaster in Chicago.* Chicago, University of Chicago Press, 2002.

64. *Improving public health responses to extreme weather/heat-waves.* EuroHEAT, 2009.

65. Mendell MJ, Smith AH. Consistent pattern of elevated symptoms in air-conditioned office buildings: a reanalysis of epidemiologic studies. *American Journal of Public Health*, 1990, 80:1193–1199.

66. Seppänen OA, Fisk WJ. Association of ventilation system type with SBS symptoms in office workers. *Indoor Air*, 2002, 12(2):98–112.

67. Flannigan B, Morey P. *Control of moisture problems affecting biological indoor air quality. ISIAQ guideline.* Ottawa, International Society of Indoor Air and Climate, 1996.

68. Cunliffe D et al., eds. *Water safety in buildings.* Geneva, World Health Organization, 2011.

69. Bartram J et al., eds. *Legionella and the prevention of legionellosis.* Geneva, World Health Organization, 2007.

70. Sieber WK et al. The National Institute for Occupational Safety and Health indoor environmental evaluation experience: Part three – associations between environmental factors and self-reported health conditions. *Applied Occupational and Environmental Hygiene*, 1996, 11:1387–1392.

71. Seppänen OA, Fisk WJ, Mendell MJ. Association of ventilation rates and CO_2 concentrations with health and other responses in commercial and institutional buildings. *Indoor Air*, 1999, 9:226–252.

72. Mendell MJ et al. Environmental risk factors and work-related lower respiratory symptoms in 80 office buildings: an exploratory analysis of NIOSH data. *American Journal of Industrial Medicine*, 2003, 43:630–641.

73. Mendell MJ et al. Indicators of moisture and ventilation system contamination in US office buildings as risk factors for respiratory and mucous membrane symptoms: analyses of the EPA BASE data. *Journal of Occupational and Environmental Hygiene*, 2006, 3:225–233.

74. Mendell MJ et al. Risk factors in heating, ventilating, and air-conditioning systems for occupant symptoms in US office buildings: the US EPA BASE study. *Indoor Air*, 2008, 18:301–316.

75. Liu L et al. The effect of air-conditioning parameters and deposition dust on microbial growth in supply air ducts. *Energy and Buildings*, 2010, 42 (4): 449–454.

76. Weschler C. Ozone's impact on public health: contributions from indoor exposures to ozone and products of ozone-initiated chemistry. *Environmental Health Perspectives*, 2006, 114:1489–1496.

77. Zuraimi MS et al. Home air-conditioning, traffic exposure, and asthma and allergic symptoms among preschool children. *Pediatric Allergy and Immunology*, 2010.

78. Menne B et al. Indoor heat protection measures and human health. In: Matthies F, Menne B, eds. *Preparedness and response to heat-waves in Europe, from evidence to action: Public health response to extreme weather events.* Copenhagen, WHO Regional Office for Europe, in press.

79. Payne A, Duke R, Williams RH. Accelerating residential PV expansion: supply analysis for competitive electricity markets. *Energy Policy*, 2001, 29:787–800.

80. Gutschner M et al. *Potential for building integrated photovoltaics.* Paris, International Energy Agency, 2001.

81. Mattick K, Durham K, Hendrix M, Slader J, Griffith C, Sen M, Humphrey T, 2003. The microbiological quality of washing-up water and the environment in domestic and commercial kitchens. Journal of Applied Microbiology 94: 842–848.

82. Oswald W et al. Fecal Contamination of Drinking Water within Peri-Urban Households, Lima, Peru American. Journal of Tropical Medicine and Hygiene, 2007, 77(4): 699–704.

83. Pokhrel AK et al. Tuberculosis and indoor biomass and kerosene use in Nepal: a case control study. *Environmental Health Perspectives*, 2010, 118(4):558–564.

84. Fthenakis VM, Moskowitz PD. Photovoltaics: environmental, safety and health issues and perspectives. *Progress in Photovoltaics*, 2000, 8:27–38.

85. *A human health perspective on climate change.* Environmental Health Perspectives and National Institute of Environmental Health Sciences, 2010.

86. Sumner S, Layde P. Expansion of renewable energy industries and implications for occupational health. *Journal of the American Medical Association*, 2009, 302(7):787–789.

87. *Sources and effects of ionizing radiation. UNSCEAR 2008 Report. Vol. II.* United Nations Scientific Committee on the Effects of Atomic Radiation UNSCEAR 2008 Report to the General Assembly, with Scientific Annexes C, D and E.

88. *Light's labours lost: policies for energy-efficient lighting.* Paris, International Energy Agency, 2006.

89. *British standards 8206. Lighting for buildings. Code of practice for daylighting.* London, BSI, 2008.

90. *Review of health and safety risk drivers.* London, Communities and Local Government, 2007.

91. Davidson M. Perception of safety and fear of crime. In: Ormandy D, ed. *Housing and health in Europe: the WHO LARES project.* Oxon/New York, Routledge, 2009.

92. Haynes PL, Ancoli-Israel S, McQuiad J. Illuminating the impact of habitual behaviors in depression. *Chronobiology International*, 2005, 22:279–297.

93. Kawamoto K et al. *Electricity used by office equipment and network equipment in the US.* Berkeley, Lawrence Berkeley National Laboratory, 2001.

94. Roth KW, Goldstein F, Kleinman J. *Energy consumption by office and telecommunications equipment in commercial buildings. Volume 1: energy consumption baseline.* Cambridge MA, Arthur D Little, 2002.

95. Household appliances, consumer electronics and office equipment. *Climate change 2007: working group III: mitigation of climate change.* The Intergovernmental Panel on Climate Change. (http://www.ipcc.ch/publications_and_data/ar4/wg3/en/ch6s6-4-11.html#footnote8#footnote8)

96. *30 years of energy use in IEA countries.* Paris, IEA, 2004.

97. *China energy databook 2004, version 6.* Berkeley, Lawrence Berkeley National Laboratory, 2006.

98. Gilbertson J, Green G, Ormandy D. *Decent homes, better health. Sheffield "Decent Homes" health impact assessment.* Sheffield, Sheffield Hallam University, 2006.

Notes

4

Shanghai, China: Urban land use and transport have profound impacts on the health of the residential housing environment as well as the carbon footprint.

Philippa Howden-Chapman

Gap analysis: Optimizing health benefits and correcting risks of mitigation strategies

Clearly many potential health co-benefits are associated with the mitigation measures reviewed in Chapter 3. This is apparent from the literature review, despite its limitations. Many measures also have a high level of cost-effectiveness and can affect multiple health outcomes, sometimes simultaneously.[1] Moreover, cost-benefit analysis shows concrete and immediate health savings in terms of reduced mortality and illness and fewer doctor visits and hospitalizations. These can help drive mitigation policies, particularly when the economic value of emissions savings are relatively lower and more long-term.

Along with those benefits, a number of potential risks were identified that also need to be better understood and addressed to fine-tune strategies that are truly win–win. For instance, thermal insulation can only be beneficial to health if used with sufficient ventilation to avoid increased indoor air pollution.

This chapter undertakes a "gap analysis" to explore where health and mitigation strategies can be enhanced or fine-tuned, and where new and complementary strategies can be proposed for mutual benefit, examining in turn:

1. Health co-benefits and risks of IPCC-reviewed mitigation measures; and
2. Neglected co-benefit opportunities around the following issues:
   - Healthy urban design – how "eco-design" can support health and mitigation more effectively;
   - Addressing health inequalities in mitigation – how low-carbon strategies can be adapted to poor countries to promote health;
   - Occupational factors – risks and exposures to construction workers and from construction materials; and
   - Behavioural change – factors that may promote or confound strategies.

4.1 Health co-benefits and risks of IPCC-reviewed mitigation measures

While incomplete, the health literature reviewed in Chapter 3 identifies a broad range of co-benefits that can be attached to these policies. These are summed up more comprehensively in Table 7 (p. 81–82). Ratings are given in terms of the potential of the mitigation strategy to generate health risks or provide health co-benefits as follows:

-- (strongly negative health impact); - (negative health impact);
+ (positive health impact); ++ (strongly positive health impact).

These ratings are based on qualitative evaluation with expert input and peer review, and with reference to the number of studies available, study design and sample size, and degree of potential confounding factors. This should be regarded as indicative rather than definitive. Effects on health equity were considered challenging to quantify in this way, so were addressed instead with comments.

4.2 Neglected co-benefit opportunities: healthy urban design

Pedestrians in a city square of Wellington, New Zealand.
(Photo: Harry Chapman)

It is estimated that 60% of the global population will live in cities by 2030, greatly increasing the total human population exposed to extreme heat. While some urban design features (clustered buildings) are briefly noted in the IPCC review as a means to reduce cooling loads, they deserve far greater attention in conjunction with housing strategies that have co-benefits for health and GHG emission reductions.

A growing body of literature points to the many synergistic health benefits that can be obtained through compact urban planning. Such planning can permit multiple carbon efficiencies in housing, along with facilitating ease of mobility, especially among children and the elderly. Replacement of wide car lanes with space-efficient designated public transport corridors and biking/walking routes can provide green space. Better use of sustainable urban design principles can thus reduce the urban heat island effect and promote multiple health co-benefits, e.g. from increased active travel, reduced exposure to outdoor air pollution and reduced ambient air pollution exposures due to private car use. Key points are summarized below.

4.2.1 Housing densities and building height

In cities and towns, clustered mid-rise housing densities are often associated with friendlier pedestrian environments and safer independent mobility of children and older adults, as well as higher carbon efficiencies than either very high- or low-density housing.

As compared to very low-density (single-family) homes, clustered apartments have lower home heating costs since walls are shared. Mid-rise densities also may often be the most "child-friendly," permitting children to safely leave their homes to play and walk to school and making neighborhoods attractive to pedestrians.[2,3] This can have important developmental health benefits for children who "learn by moving," as well as for the elderly who might otherwise experience mobility restrictions. Further study is needed on correlations between densities and independent child mobility.

While housing styles are also a factor of geography and available land space, very high-rise housing may have a range of negative health impacts in many settings. Some of those same health impacts vary widely, as they may be mitigated or exacerbated by culture, socioeconomic status and social factors. Nonetheless, high-rise housing inevitably requires long staircases or elevators for access. This, together with much higher population densities in the immediate home and "neighborhood" environment, would tend to limit children's independent mobility in and around their immediate neighborhood environment. High-rise housing and high housing densities, as well as very low-rise

Table 7. Climate change and health co-benefits of IPCC-reviewed strategies: summary table

Mitigation strategy	Likely health co-benefits	Impact of health co-benefit	Health risks to be avoided	Impact of health risk
Improved thermal performance of building envelope (IPCC 6.4.2)	**Environmental exposure** Thermal comfort Noise exposure reduction	 ++ +	Risk of inadequate ventilation: a) Reduced indoor air quality leading to potentially increased concentrations of indoor air pollutants (e.g. radon, mould and moisture) as a cause of asthma, bronchial obstruction and other illnesses b) Increased airborne infection transmissions (e.g. TB); risk of exposure to health-damaging insulation materials and fibres that cause cancer and other illnesses	- - - -
	Disease risk reduction Reduced cardiovascular diseases, bronchial obstruction, asthma and other respiratory conditions Reduced vector-borne disease due to infestations and pests Better mental health through thermal comfort	 ++ ++ +		
	Equity impacts Depends on access of poor to improvements	 +		
Low-carbon-emissions heating systems and passive solar design (IPCC 6.4.3, 6.4.6–7)	**Environmental exposure** Thermal comfort Hygiene	 ++ +	Field studies have found that more cost- and energy-efficient heating do not always reduce net household energy use (and thus energy-related greenhouse gases and air pollutants) by an equivalent amount. This is because some households may allocate a portion of their cost savings to *increase* their energy (electricity or heat) consumption, a phenomenon described as the "take-back effect"	 0
	Disease risk reduction Reduced asthma and respiratory symptoms related to cold exposure, damp and mould Reduced pneumonia and COPD (in case of reduced biomass use) Better mental health due to better thermal comfort	 ++ ++ +		
	Equity impacts Depends on access of poor to improvements	 +		
Reduced cooling loads on buildings through design features and improved natural ventilation (IPCC 6.4.4)	**Environmental exposure** Thermal comfort	 ++	May not work when night temperatures remain high; need to be adapted to regional humidity Design must take account of winter as well as summer risks Natural ventilation without house screening may increase vulnerability to vector-borne diseases May increase exposure to high outdoor air pollution concentrations, causing respiratory symptoms, unless filters are used Avoid use of lead in paint (e.g. white paint for albedo effect)	0 0 - - - - -
	Disease risk reduction Reduced asthma/respiratory illness from particulates, radon, mould, etc. Reduced TB and other airborne infection transmission risk Less airborne disease transmission via air-conditioning systems	 ++ ++ +		
	Equity impacts High equity co-benefit from broader access to effective cooling and ventilation, particularly when design measures are adopted in low-income settings	 +		

Strongly positive health impact ++; Positive health impact +; Strongly negative health impact: - - ; Negative health impact: -

Mitigation strategy	Likely health co-benefits	Impact of health co-benefit	Health risks to be avoided	Impact of health risk
More energy-efficient and better-maintained heating, ventilation and air conditioning systems (HVAC) Greater reliance on building design and natural ventilation (IPCC 6.4.4–5)	**Environmental exposure** Thermal comfort Reduced noise exposure	++ +	Greater risk of airborne infectious diseases (e.g. tuberculosis) and upper and lower respiratory symptoms in AC rooms/spaces lacking sufficient fresh air exchanges	- -
	Disease risk reduction In settings with significant outdoor air pollution, reduced respiratory symptoms and asthma Less risk of cardiovascular disease due to heat exposure Less risk of vector-borne disease due to closed windows	++ ++ +	Increased urban dependence on AC stimulates vicious cycle of exacerbated urban heat island effect More noise and pollution exposure for those not using air conditioning Bacterial proliferation/legionellosis in very large HVAC tanks/cooling towers Delayed climate-related health impacts from added greenhouse gas emissions of air conditioners	- - - - -
	Equity impacts Those least able to afford AC suffer the most from its noise and heat island impacts.	-		
Passive solar hot water and photovoltaic solar electricity (IPCC 6.4.7–8)	**Environmental exposure** Hygiene and sanitation	+	Greater initial cost outlays pose barriers for poor families if not offset by subsidies	-
	Disease risk reduction Less asthma and respiratory disease due to decreased use of kerosene lighting in developing countries Fewer burns from kerosene appliances	+ +	New technology risks require more assessment, including of occupational and environmental risks of production and exposure to waste byproducts, e.g. respiratory irritations and impacts of exposures to heavy metals or other toxic substances.	0
	Equity impacts More access to electricity among poor and rural populations Lower long-term electricity cost once initial investment is made	++ +		
Lighting and day lighting: window positioning to reduce heat/cold impacts; highly energy-efficient indoor lighting (IPCC 6.4.9–10)	**Environmental exposure** Thermal comfort	++	Household injury from inadequate indoor/proximity lighting	-
	Disease risk reduction Less asthma and respiratory disease due to natural ventilation through windows Fewer home injuries (falls) Positive effect of light on metabolic function and mental health	+ ++ +		
Household appliances and electronics: more low-energy and direct-current appliances, including improved biomass cookstoves (IPCC 6.4.11; 6.6.2)	**Environmental exposure** Reduced indoor air pollution Improved food safety, kitchen hygiene	++ +	Equity gains dependent on increased access of poor to new low-energy cookstove technologies and other appliances	-
	Disease risk reduction Reduced asthma and respiratory disease Fewer injuries from burns due to inadequate cooking and heating appliances Less COPD, cancer and cardiovascular disease	+ ++ +	In developed countries, more efficient appliances may not decrease GHG and air pollution emissions if there is not a equivalent decrease in overall energy use	-
	Equity impacts Access to cleaner biomass and biogas cookstoves	++		

housing, may also lead to anonymity and a decreased sense of safety in the presence of not enough or, conversely, too many "strangers" in and around the immediate home environment.

Building heights also have a major impact on ventilation and cooling factors. A specific discussion of the airflow in wind tunnels created by high-rises is provided by Santamouris.[4] Generally, the lower the height-to-width ratio of built space, the better the penetration of air. In very dense environments where the height-width ratios are very high, and when the wind direction is perpendicular or oblique to the canyon axis, the air flow may instead be governed by thermal forces. A coupling of wind flow above the buildings and the air flow inside the wind tunnel can lead to high undisturbed wind speeds (> 5 m/sec).[5]

4.2.2 Connectivity to dedicated public transport, walking and cycling

Per passenger, a fully-occupied car may consume more than 2–3 times the road space of a fully-occupied bus or light rail car travelling at similar speeds.[6] Thus development of dedicated public transport lanes can free up valuable urban space for walking/cycling infrastructure and parks. This, in turn, helps reduce ambient air pollution exposures and promotes active travel and better cardiovascular health. Public transport, walking and cycling are typically most effective in compact urban areas that reduce travel distances and concentrate travel destinations, reinforcing other health and environment synergies. Research on health co-benefits of such strategies is summarized in the *Health in the Green Economy* report on the co-benefits to health of climate change mitigation in the transport sector.

> Per passenger, a fully-occupied car may consume more than 2–3 times the road space of a fully-occupied bus or light rail car traveling at similar speeds. Thus development of dedicated public transport frees valuable urban space for walking/cycling and parks.

Residents of streets with relatively lower traffic volumes tend to report a greater sense of connectedness and more positive links with their neighbors than those living in areas with heavy traffic.[7] Levels of air pollution and traffic congestion, safety from injury, security from crime and ease of movement around the housing environment and neighborhood are important factors in housing valuation, as well as in more subjective feelings of well being.[8]

4.2.3 Urban landscaping and traditional design principles

Urban landscaping and adaptation of traditional design principles may include tree planting, green spaces and ponds or water fountains in courtyards. These natural cooling modes encourage natural ventilation, thermal comfort and sanitation as well as mental health. Such principles are often embedded in traditional urban and building design principles of many cultures and have multiple health co-benefits. Indigenous knowledge of ways to keep dwellings cool and/or warm can offer valuable and energy-efficient solutions. An analysis of the thermal performance of Chinese traditional vernacular dwellings in the Wannan area shows a successful cooling effect due to consideration of: sun, shading and insulation; selection of building materials for envelope and roof; and thermally-sensitive design of rooms and courtyards.[9]

Trees

Protection from solar radiation using trees reduces urban temperatures overall, and tree cover has a large direct physiological effect in reducing heat stress, especially for pedestrians.[10–12] Trees create a favourable thermal balance for humans and enhance outdoor thermal comfort.[13,5] Trees also absorb sound, reduce noise stress, produce oxygen, filter particulate pollutants and reduce wind speeds, reducing air pollution exposures as well as exposures to unhealthy wind velocities. Risks of respiratory and cardiovascular disease and stroke may be reduced due to the filtration of ambient pollutants by tree cover.[14] The planting of food/fruit trees for shade also enhances food security.

Parks

Urban parks offer essential filtering of air pollution and shade that help reduce heat island impacts. (Photo: istockphoto)

Parks reduce heat stress and air pollution exposures while providing opportunities for physical activity and active travel, improving well being and mental health.[i,15,16,17] Evidence across a range of sources suggests that contact with safe green spaces can improve outcomes for a range of public health and social indicators. Having green spaces in an area can reduce health inequalities, improve well being and aid in treatment of mental illness. Some analysis suggests that physical activity and contact with nature can help remedy mild depression and reduce physiological stress indicators.[18,19] Parks help keep a city cool during heat waves, although high ozone concentrations that often accompany heat waves also are damaging for plants, which also filter polllution and are otherwise so important to urban air quality.[20]

Water fountains, ponds or lakes

In health terms, these can also contribute directly to increased thermal comfort and, if well-maintained, also sanitation (e.g. possibilities for face- and hand-washing and bathing). Fountains, ponds and lakes are also a source of cultural tradition and inspiration. In many parts of India, for instance, traditional rainwater harvesting systems are important local heritage spots. In environmental terms, these water sources decrease air temperatures via convection and evaporation. Ponds and fountains cool open spaces because water temperatures increase more slowly than air temperatures. The cooling capacity of ponds depends on their depth, number and nature of sprays, operational schedules and whether the pond is shaded. Santamouris et al. provide a full analysis of various techniques to use water in outdoor spaces.[20] Water also cools down more slowly than air, and so in prolonged periods of hot weather, cooling by water loses efficacy. Heat-sensitive images of London during the August 2003 heat wave thus showed especially air warm temperatures over water bodies.

---

i According to von Stülpnagel and von Stülpnagel et al., urban parks reduce air temperatures in the adjacent neighbourhoods; however, this effect was limited to a relatively small zone extending only 200–400 m from the margin of a large park on a calm day. Von Stülpnagel A. *Klimatische Veränderungen in Ballungsgebieten unter besonderer Berücksichtigung der Ausgleichswirkung von Grünflächen, dargestellt am Beispiel von Berlin-West.* (Climatic changes in urban areas with special attention to the compensatory effect of green areas, the example of West Berlin). Diss. am Fachbereich 14 (Dissertation in the section 14). Berlin, Technische Universität, 1987; also von Stülpnagel A et al. The importance of vegetation for the urban climate. In: Sukopp H, ed. *Urban Ecology*. The Hague, SPB Academic Publishing, 1990.

Building courtyards

Along with their function in cooling and shading, building courtyards may provide safe areas for physical activity, particularly by children, offering another dual health and mitigation benefit.

4.3 Addressing housing and health inequities with mitigation strategies

As noted, developing cities are the fastest-growing on the planet. Nearly 40% of such growth is in slums, while at the other extreme energy-intensive high-rises and gated suburban communities house the affluent of developing as well as developed countries. This represents a huge and growing housing equity and health gap.

Innovative low-carbon solutions could help close some of this gap using measures that also address basic and vital health needs more systematically. Key "co-beneficial" approaches that require further research and evaluation in health and mitigation literature include the following:

- Redevelopment of poor housing with low-carbon designs more resilient to extreme heat, cold, rain, storms and drought.
- Low-carbon household energy solutions that also improve indoor and outdoor air quality.
- Equitable access to low-carbon public transport and walking and cycling networks that generate multiple health benefits in terms of air pollution exposures, injuries and physical activity.
- Equitable access to safe drinking-water, improved sanitation and waste disposal through grid expansion or small community water harvesting systems. Along with immediate reductions in risk of diarrhoea and other diseases, these provide long-term savings in terms of water extraction costs, water waste and sanitation treatment. Early provision of services in the housing development phase can also promote more "compact" urban development by clustering housing around services.
- Solar photovoltaic electricity, lights and appliances can immediately reduce risks of respiratory illnesses, eye conditions and injuries as well as long-term carbon savings in terms of reduced fuel use (see Chapter 6 case study on solar lanterns).
- Expanded access to hot water through solar hot water heating. In developed countries, this measure may not make significant differences in household health and hygiene insofar as hot water is already available. However, studies on water, sanitation and hygiene show that increasing access to hot water where it was unavailable before can reduce bacterial loads, self-reported respiratory conditions and eye problems, and enable easier bathing and clothes-washing.[21]
- Better house screening, natural ventilation and household water management reduce risks of TB and vector-borne disease transmission as well as providing long-term carbon efficiencies.

Biogas latrine built as part of a clean fuel project undertaken by "the World in Justice" NGO in a village of rural Nepal. The health benefit is three-fold: improved access to sanitation through the latrine hookup, reduced indoor air pollution from the use of biogas for cooking; and reduced climate change through use of a renewable fuel source. (Photo: Heather Adair-Rohani)

4.3.1 Slum housing upgrade with pedestrianization

Housing upgrades can impact health and promote sustainability in multiple ways. In the case of a densely populated Indonesian working-class neighborhood or *kampong*, traffic-choked alleys were reclaimed as pedestrian enclaves and "greened" with pocket gardens as part of a programme of water and sewage improvements and housing rehabilitation. This can help reduce air pollution exposure, prevent injury, improve child health and promote active transport.

4.3.2 House screening for natural ventilation and vector-borne disease control

In malaria-endemic regions, the only way to safely promote natural night ventilation is in combination with mosquito protection. As with bednets, substantial evidence finds window/door screens a highly effective, low-carbon-footprint housing measure, particularly when combined with other appropriate design and water management features. Initiatives to promote such low-carbon housing improvements can potentially improve health as well as reduce reliance on vector-control chemicals and (where it is affordable) air conditioning.

A 2005 systematic literature review by Keiser et al.[22] identified 40 studies that examined the efficacy of housing modification and environmental modification or manipulation against clinical malaria outcomes (see excerpt, Table 8). Most of the studies (85%) were implemented before the Global Malaria Eradication Campaign (1955–1969), which mainly relied on indoor residual spraying with dichlorodiphenyltrichloroethane (DDT).

In eight housing studies mostly related to window/door screening, the risk ratio of malaria was reduced by 79.5% (95% CI 67.4–87.2). In 16 studies on environmental modification, the risk ratio of malaria was reduced by 88% (95% CI 81.7–92.1).[22] Some of the most striking successes were recorded in highly endemic malaria regions such as Zambia, where combined window/door screening and water and environmental management/modification reduced malaria incidence by 50%–75% in the first 3 to 5 years of the control programme, which was prior to DDT use.[23]

The authors conclude that "a negative aspect of the Global Malaria Eradication Campaign was that during and after the campaign little attention had been paid to the relatively straightforward task of modifying human habitation." We are not aware of any recent malaria control programmes that have used human habitation modification in a systematic manner to reduce malaria morbidity and mortality.

Screening and other housing design features such as closed eaves and ceiling design[24] may be important components in design of healthy low-carbon housing interventions in malaria-endemic countries. By promoting better natural ventilation, these measures help reduce transmission of airborne respiratory illnesses as well.[25,26] Studies from the systematic review are described in Table 8.

Schoolgirls stroll in the lane of an Indonesian neighborhood recently reclaimed for pedestrians. (Photo: Jeff Kenworthy)

Table 8. House screening to prevent malaria: health outcomes in a systematic review

	Principal malaria vectors	Intervention of human habitation	Additional intervention	Clinical malaria parameters			Estimated risk ratio (95% CI)
				Number of participants	Control group	Intervention group	
Lazio, Italy 1899–04	..	Windows covered and doors screened	NA	Pontegalera line 1900 Intervention group, 36; Control group, 42	39 cases	2 cases	0.06 (0.02–0.20)
Asinara, Italy 1900	..	Windows of convicts' dormitories covered with strong muslin	Petrolising of breeding sites	..	40 cases of malaria had been contracted in Asinara in 1899	No cases of malaria was contracted in 1900	NA
Missouri, USA 1923	..	Mosquito proofing of houses	NA	Intervention group, 513; Control group, 698	Malaria incidence in non-protected houses 18×2% per year	Malaria in protected houses 8×8% per year	0.30 (0.21–0.41)
Lahore, India 1925–27	..	Mosquito proofing of British infantry barracks (wire netting, double doors)	NA	Intervention group, 285; Control group, 281	NA	Malaria incidence of British units in 1925:569 per 1000 per year; malaria incidence of British units in 1927: 45 per 1000 per year	0.08 (0.05–0.14)
Amritsar, India 1925–27	..	Mosquito proofing of British infantry barracks (wire netting, double doors)	NA	Intervention group, 137; Control group, 199	NA	Malaria incidence of British units in 1925:613 per 1000 per year; malaria incidence of British units in 1927: 58 per 1000 per year	0.01 (0.05–0.18)
Missouri, USA 1926	..	Mosquito proofing in open and closed houses	NA	Closed houses: Intervention group, 846; Control group: 246 Open houses: Intervention group, 408; Control group, 258	Malaria incidence in non-protected closed houses 19×1% per year; malaria incidence in non-protected open houses 23.5% per year	Malaria incidence in screened closed houses 5×1% per year; Malaria incidence In screened open houses 12×7% per year	Closed houses 0.27 (0.18–0.40); Open houses 0.54 (0.38–0.75)
Honduras, 1926	..	Mosquito proofing of houses	NA	Intervention group, 135; Control group, 2607	Malaria incidence in non-protected houses 29×1% per year	Malaria incidence in protected houses 6×6% per year	0.23 (0.12–0.42)

	Principal malaria vectors	Intervention of human habitation	Additional intervention	Clinical malaria parameters			Estimated risk ratio (95% CI)
				Number of participants	Control group	Intervention group	
Leflore, Missouri, USA 1927	..	Screened houses	NA	Intervention group, 104; Control group, 104	84 malaria cases in non-protected houses	24 malaria cases in screened houses	0.28 (0.20–0.41)
South Africa 1930–31	*A costalis* *A funestus*	Screened houses	NA	..	NA	Malaria incidence in children reduced by 50%	NA

Source: Keiser J et al, 2005

There is also substantial experience with use of habitation improvements to protect against other vector-borne diseases, including Chagas disease and dengue, as well as initial evidence on housing measures that might be useful in visceral leishmaniasis control. Case studies are provided in Chapter 6.

4.4 Occupational health – risks and exposures to construction and building renovation workers

Significant occpuational hazards are associated with construction, construction products and also with housing retrofits that are integral to many energy efficiency initiatives. Many of these risks can be avoided by simple occupational health measures. These, are not addressed by the IPCC asssessment, and they deserve further analysis.

In the climate change context, removal of older kinds of insulation and building materials, may involve significant exposures to potentially health-damaging materials as well as to construction dust. Similarly, occupational health must be considered when developing, introducing and installing building and construction materials that may have unknown health risks. There may be unknown occupational risks associated with the use of any new product, and those risks may extend to materials or technologies used in what is regarded as "green construction" as well as more traditional practices. (See Fig. 7).[27] Along training and informing construction workers about good practice and protective measures, emphasis on good occupational health in the construction phase can also yield a "spill-over" impact for both manufacturers and householders. Increased awareness of occupational health hazards of asbestos removal, for example, can increase awareness of its health hazards as a construction material more generally.

Fig. 7. Summary of occupational hazards in green construction

Increased risks from existing hazards

Skylights: falls

Atriums: falls, ergonomics

Recycling: strains, sprains and punctures; slips and falls; "struck-by" hazards

Recycled materials: coal ash in concrete

Weatherization: lead and asbestos exposures, electrical

Indoor air quality: heat stress

Hazards associated with new technologies and products

Solar power: falls, electrical, ergonomics, burns, exposure to toxins

Wind power: falls, electrical

Weatherization: exposure to isocyanate and silica

Building materials: exposure to silica and nanomaterials

Source: *Green and Healthy Jobs.*[27]

4.5 Behavioural change: factors that promote or confound strategies

Building user behaviour can have a major impact both on GHG emissions and on health. Along with IPCC reference to better air conditioning control (IPCC 6.4.5.), behaviour, as such, is addressed only briefly in an analysis relating to behavioural barriers (IPCC 6.7.6):

> "The potential impact of lifestyle and tradition on energy use is most easily seen by cross-country comparisons. For example, dishwasher usage was 21% of residential energy use in UK residences in 1998 but 51% in Sweden... Cold water is traditionally used for clothes washing in China … whereas hot water washing is common in Europe. Similarly, there are substantial differences among countries in how lighting is used at night, room temperatures considered comfortable, preferred temperatures of food or drink, the operating hours of commercial buildings, the size and

User knowledge about proper operation and maintance of new technologies, such as PV solar lighting, is important to their successful introduction.

composition of households, etc. ... Variation across countries in quantity of energy used per capita, which is large both at economy and household levels ... can be explained only partly by weather and wealth; this is also appropriately attributed to different lifestyles."

Both extremely hot and extremely cold homes can be unhealthy. In *The Lancet* review of health co-benefits of different energy efficiency strategies by Wilkinson *et al.* models how outdoor air pollution would be reduced in the UK as a result of home energy saving strategies, including temperature reduction in homes heated over 18°C by just 1°C. [28] The research concludes that a combined approach including thermostat control as well as thermal envelope improvements and fuel switching would more significantly reduce fossil-fuel generated outdoor air pollution, and increase health gains per megatonne of CO_2 saved. Behavioural change can thus contribute to climate change mitigation, the paper concludes.[ii]

Especially in residential buildings, appropriate ventilation is also partly or largely a consequence of behaviour, insofar as opening windows is an effective measure for ventilating indoor environments, including for brief periods in the winter. [29]

There needs, however, to be knowledge about when and how best to ventilate effectively, and how to do so in a way that avoids overheating or overcooling of indoor spaces. For instance, closing windows and shutters during the day and opening them to cooler evening air is a traditional and effective practice in many Mediterranean countries that supports natural cooling and night ventilation.[5]

However, a range of objective external factors that can severely limit householders' ability to ventilate adequately. These may include outdoor noise and security issues prevalent in many cities, as well as, in many hot climates, fear of vector-borne disease.

Other behavioural measures that may improve thermal comfort and thus reduce the need for mechanical ventilation and air conditioning include: adjusting clothing to weather conditions and indoor temperatures; correct use and maintenance of heating and cooling appliances also contributes both to energy efficiency and reduce risks associated with mechanical heating and cooling systems.

Behavioural measures related to protective clothing for heat, dampness, cold, exercise and physical activity, diet and personal hydration are other important factors that moderate the physiological response to extremes of heat and cold, indoors as well as outdoors.

For instance, ample personal hydration and protective dress (e.g. hats and head covers) in hot weather are well-known ways to reduce heat stress in hot climates; these are now being emphasized more in temperate regions such as Europe as part of heat wave response. Additionally, most cold weather cultures have well-developed traditional protective clothing and bedding measures, some of which offer excellent protection from extremes of cold. This traditional knowledge can be explored further in light of current studies about physiological response to cold.

In summary, measures to enhance buildings' energy efficiency and to reduce GHG emissions need to be supported by public health programmes with awareness of co-benefits in regards to climate change mitigation as well as adaptation strategies.

If, however, behavioural change is to be targeted as a measure to achieve linked health and mitigation objectives, evidence shows that a staged and multi-faceted approach is more effective. This typically includes: information provision, goal setting, commitment and feedback. In this manner, rather than promoting just one measure alone, different behavioural barriers may be addressed to achieve real and lasting change.[iii,5,30,31]

---

ii The research model hypothesizes that health impacts of lowering thermostats would be negligible, noting that average indoor winter-time temperatures in the UK have risen substantially since the 1970s. However, the authors caution that results must be interpreted cautiously due to scarce evidence about actual temperature thresholds for cold among different population subsets, as well as feasibility of such measures. In reality, temperature ranges in homes would be broader, and impacts of personal choices more complex.

iii The (summer) temperature limits are primarily based on studies in office buildings. Nevertheless, based on general knowledge of thermal comfort and human responses, the limits may be assumed to apply to comparable buildings with mainly sedentary activities, such as residential buildings.

References

1. *Pathways to a low-carbon economy: version 2 of the global greenhouse gas abatement cost curve.* McKinsey & Co., 2009.

2. Churchman A. Children in urban environments, the Israeli experience. In: Michelson E, Michelson W, eds. *Managing urban space in the interest of children.* Canada, UNESCO Programme on Man and Biosphere Committee, 1980:51.

3. Churchman A. The urban environment and women in Israel. In: *Habitat II shadow report*, UNESCO Programme on Man and Biosphere, 1996:22–53.

4. Santamouris M, ed. Energy in the urban built environment. London, James and James, 2001.

5. Menne B et al. Indoor heat protection measures and human health. In: Matthies F, Menne B, eds. *Preparedness and response to heat-waves in Europe, from evidence to action. Public health response to extreme weather events.* Copenhagen, WHO Regional Office for Europe, in press.

6. *Policy briefing on transport, environment and health in developing cities.* Geneva, World Health Organization, 2010. (http://www.who.int/heli/risks/urban/en/)

7. Appleyard D. *Livable Streets.* Berkeley, University of California Press, 1981.

8. *Transport, environment and health in developing cities.* UNEP-WHO Health and Environment Linkages Initiative. (http://www.who.int/heli)

9. Borong L et al. Study on the thermal performance of Chinese traditional vernacular dwellings in summer. *Energy and Buildings,* 2004, 36:73–79.

10. Harrison C et al. *Accessible natural green space in towns and cities: a review of appropriate size and distance criteria.* Peterborough, English Nature Research Reports No. 153, 1995.

11. Eliasson I. The use of climate knowledge in urban planning. *Landscape and Urban Planning,* 2000, 48:31–44.

12. Handley J et al. *Accessible natural green space standards in towns and cities: a review and toolkit for their implementation.* Peterborough, English Nature Research Reports No. 526, 2003.

13. Picot X. Thermal comfort in urban spaces: impact of vegetation growth. Case study: Piazza della Scienza, Milan, Italy. *Energy and Buildings,* 2003, 36:329–334.

14. *A human health perspective on climate change.* Environmental Health Perspectives and the National Institute of Environmental Health Sciences, 2010.

15. Guite HF, Clark C, Ackrill G. The impact of the physical and urban environment on mental well-being. *Public Health,* 2006,120(12):1117–1126.

16. Von Stülpnagel A. *Klimatische Veränderungen in Ballungsgebieten unter besonderer Berücksichtigung der Ausgleichswirkung von Grünflächen, dargestellt am Beispiel von Berlin-West.* [Climatic changes in urban areas with special attention to the compensatory effect of green areas, the example of West Berlin]. Diss. am Fachbereich 14 [dissertation in section 14]. Berlin, Technische Universität, 1987.

17. Von Stülpnagel A et al. The importance of vegetation for the urban climate. In: Sukopp H ed. *Urban Ecology.* The Hague, SPB Academic Publishing, 1990.

18. Kirsch I et al. Initial severity and antidepressant benefits: a meta-analysis of data submitted to the Food and Drug Administration. *PLoS Medicine,* 2008, 5(2):e45.

19. Bird W. *Natural thinking: investigating the links between the natural environment, biodiversity and mental health.* England, Royal Society for the Protection of Birds, 2007. (http://www.rspb.org.uk/Images/naturalthinking_tcm9-161856.pdf)

20. Santamouris et al. *Cooling the cities.* Paris, Eyrolles, 2004.

21. Christen A, Navarro CM, Mäusezahl D. Safe drinking water and clean air: an experimental study evaluating the concept of combining household water treatment and indoor air improvement using the water disinfection stove (WADIS). *International Journal of Hygiene and Environmental Health* 2009, 212(5):5628.

22. Keiser J, Singer B, Utzinger J. Reducing the burden of malaria in different eco-epidemiological settings with environmental management: a systematic review. *Lancet Infectious Diseases,* 2005, 5(11):695–708.

23. Utzinger J, Tozan Y, Singer BH. Efficacy and cost-effectiveness of environmental management for malaria control. *Tropical Medicine and International Health,* 2001, 6(9):677–687.

24. Lindsay SW, Emerson PM, Charlwood JD. Reducing malaria by mosquito-proofing houses. *Trends in Parasitology,* 2002, 18:510–514.

25. Wolff CG, Schroeder DG, Young MW. Effect of improved housing on illness in children under 5 years old in northern Malawi: a cross-sectional study. *British Medical Journal,* 2001, 322:1209–1212.

26. Lindsay SW et al. Changes in house design reduce exposure to malaria mosquitoes. *Tropical Medicine and International Health,* 2003, 8(6):512–517.

27. Chen H. *Green and healthy jobs.* Silver Spring, Center for Construction Research and Training, 2010.

28. Wilkinson P et al. Public health benefits of strategies to reduce greenhouse-gas emissions: household energy. *Lancet,* 2009; published online Nov 25. DOI:10.1016/S0140-6736(09)61713-X.

29. Fengfeng Z et al. Study on ensuring indoor fresh air volume in winter. *Measurement,* 2010, 43(3): 406–409.

30. Abrahamse W et al. A review of intervention studies aimed at household energy conservation. *Journal of Environmental Psychology,* 2005, 25(3):273–291.

31. Abrahamse W et al. The effect of tailored information, goal setting and tailored feedback on household energy use, energy-related behaviours and behavioural antecedents. *Journal of Environmental Psychology,* 2007, 27(4):265–76.

5

Matthias Braubach

Freiburg, Germany: Mid-rise apartments surrounded by gardens, green spaces and walking paths.

Tools to assess, plan and finance healthy and climate-friendly housing

A broad range of tools exists to assess, plan and finance healthy, climate-friendly housing. This discussion briefly examines these tools and their most relevant uses.

5.1 Assessment methods (HIA, SEA, EIA)

Assessment of housing quality, including health, safety and sustainability, has two broad functions:

- provide a robust basis for policy development at all levels, as well as for compliance monitoring and for research regarding quality of housing stock;
- assist house owners, renters, property managers and compliance agencies in making informed judgements about management of individual properties.

The health sector promotes health impact assessment (HIA) to evaluate health effects of novel policies and technologies at various scales. HIA has been applied to several potential climate change mitigation strategies. Given widespread uncertainty regarding the potential health impacts of certain mitigation strategies, HIA can be a valuable tool for evaluating possible health effects, especially when used in combination with other approaches to life-cycle assessment.[1] In the housing sector, HIA has been most often used in European countries, particularly in the United Kingdom.

Addressing housing issues and improving access to high-quality and affordable housing through HIA involves an array of public and private sector agencies. Additional data are needed to demonstrate HIA's impacts on health directly and indirectly through other causal factors. Strategic environmental assessment (SEA) is used in the context of development with trans-boundary impacts, and environmental impact assessment (EIA) is often used as a regulatory tool to assess major housing and commercial developments. Where planning codes and housing development are supervised by local planning councils, EIA may vary widely in rigor and content. For instance, local EIA may the impact of housing on wetlands or traffic, but not the energy efficiency or carbon footprint of planned projects. The following examples show the use of assessment tools in housing policies of developing countries:[2]

- The United Arab Emirates have adopted a framework for sustainable design, construction and operation of communities, buildings and villas using a unique assessment tool called the Estidama Pearl Rating System that is specifically tailored to the hot climate and arid environment.

- A Task Team was set up in 1998 by South Africa to develop a national policy on environmentally efficient low-cost housing and to encourage environmentally sound practices in the housing sector. A first edition of its Environmental Implementation Plan was released in 2000; however, this plan has been blamed for delays in meeting housing targets.
- Jordan carries out Environmental Impact Assessment for major buildings.
- China developed a Healthy Housing Program (HHP) in 1999 and released *Technical Essentials for the Construction of Healthy Housing* in 2001; this was updated in 2002 and 2004. The latter served as the background for the development of a framework and tool for HHP assessment combining quantitative and qualitative indicators covering housing's physical, chemical, psychological and social aspects.[3] Feedback from more than 60 national Healthy Housing Projects showed this is a useful tool in assessing the life-cycle of housing construction and that it effectively guarantees housing's healthy performance as well as developing friendly neighbourhoods and better living environments.

5.2 Intervention studies

Health experts can contribute skills and methods for active evaluation of mitigation measures' health benefits and risks.

As important as it is, the link between housing conditions and health effects constitutes only half of the knowledge needed. For example, to recognize the link between exposure to mould and increased risk of asthma does not necessarily provide information on whether or how specific methods of reducing mould exposures improve asthma status. An intervention study may have unintended consequences, or the link between a given housing condition and a given health outcome may be spurious. Better understanding is needed of housing interventions that demonstrably improve health; such understanding can lead to policies and programmes that will improve quality of life. There is an economy of scale to this approach: one intervention can address multiple hazards. For example, the replacement of a rotted handrail covered with deteriorated lead paint addresses both lead poisoning and injury prevention.

There are two types of interventions: clinical evidence and environmental or housing measurements. Each of these sources of evidence has strengths and weaknesses. Clinical evidence (or other health data, such as self-reported health) is likely to most directly measure health status. Yet many health conditions do not have adequate biomarkers, or have long time horizons before an adverse health event occurs, making clinical evidence problematic. For example, lung cancer from radon exposure may not be clinically observable for many years, yet there is good evidence that radon environmental measurements can be linked reliably to risk of lung cancer. Similarly, asthma is a complex set of symptoms for which a single reliable biomarker has yet to be identified. Thus an intervention that successfully reduces environmental exposures for which there is good evidence of a dose-response relationship may be judged successful.

5.3 Indicator systems

Indicator systems may include "green labels" and housing safety rating systems, a few of which are noted below.

A range of countries have developed labels and standards for green building or energy efficiency for buildings. Some of these are purely voluntary, while others are a product of public-private partnerships such as the Swiss *Minergie*, supported by the Swiss Confederation and its cantons. Certain Swiss tax incentives and financial subsidies are available for construction or retrofit of *Minergie*-compliant buildings.[4]

- Australia: Nabers[5] / Green Star[6]
- Brazil: AQUA[7] / LEED Brasil[8]
- Canada: LEED Canada[9] / Green Globes[10]
- China: GBAS[11]
- Finland: PromisE[12]
- France: HQE[13]
- Germany: DGNB[14] / CEPHEUS[15]
- Hong Kong: HKBEAM[16]
- India: Indian Green Building Council (IGBC)[17] / GRIHA[18]
- Israel: Israel Standard for Green Building[19,i]
- Italy: Protocollo Itaca[20] / Green Building Counsil Italia[21]
- Japan: CASBEE[22]
- Malaysia: GBI Malaysia[23]
- Mexico: LEED Mexico[24]
- Netherlands: BREEAM Netherlands[25]
- New Zealand: Green Star NZ[26]
- Philippines: Philippine Green Building Council[27]
- Portugal: Lider A[28]
- Singapore: Green Mark[29]
- South Africa: Green Star SA[30]
- Spain: VERDE
- Switzerland: Minergie[31]
- United States: LEED[32] / Living Building Challenge[33] / Green Globes[34] / Build it Green[35] / NAHB NGBS[36] / International Green Construction Code International Green Construction Code (IGCC)
- United Kingdom: BREEAM[37]
- United Arab Emirates: Estidama[38]

> Many housing rating systems and assessment tools (and certainly most green labels) consider energy efficiencies. However, assessment of the full range of health impacts, beyond basic safety, is often incomplete or absent.

Many housing rating systems and assessment tools, (and certainly most green labels), consider energy efficiencies. However, assessment of the full range of health impacts – beyond basic safety – is often incomplete or absent. Assessment of house quality that includes health, safety and sustainability can provide a more robust basis for policy development, compliance monitoring and research on the quality of housing stock. Such assessment also can assist house owners, renters, property managers and compliance

---

i In an agreement with the Ministry of Environmental Protection, the Standards Institution of Israel has committed to accelerate the upgrading of Israel's Green Building Standard (Israel Standard 5281 on Buildings with Reduced Environmental Impact) to a mandatory government regulation, with completion scheduled for March 2011.

agencies in making informed judgements about management of individual properties.[39] This is particularly true in regard to energy-saving measures and impacts on health.

The Housing Health and Safety Rating System (HHSRS) was developed in 2001 by the United Kingdom government as a replacement for housing and health fitness standards. This has the aim of grading the severity of threats in the home and of being hazard-focused, comprehensive and evidence-based. The main principle underlying the HHSRS system is that a dwelling should provide a safe and healthy environment for any potential occupant or visitor. The system adopted by the legislation also directs improvements of housing stock.

The HHSRS relies on logical evaluation of potential risks to health and safety from any deficiencies identified in a dwelling, including those related to design and construction. The hazards are arranged into four groups reflecting basic health requirements: physiological and psychological requirements and protection against infection and accidents. Under the HHSRS, energy efficiency and inadequate heating and insulation of the dwelling are considered matters of key relevance affecting likelihood and harm outcome.[40]

In New Zealand, research knowledge about the health effects of uninsulated, unheated and cold houses (as noted in Chapter 3.4–5) has been incorporated into an Assessment tool used by the New Zealand "Healthy Housing Index", a rating system based on the UK Health and Safety Rating System. This tool has been designed to identify aspects of a house which can be remediated to improve health. Use of this tool has raised awareness about the lack of insulation in most New Zealand houses and supported economic arguments for a larger investment of public money into insulation retrofits of homes.

5.4 Regulatory frameworks, including building and planning codes

Planning codes guide building siting and density as well as green spaces, transport, utility infrastructure and other urban infrastructure features. Building codes address specific construction features and are administratively managed, usually by local, country-wide or regional personnel. While there is increased reflection of sustainable "urbanism" and rural development in planning and building procedures of developed countries, this varies widely by country and is far less prevalent in developing countries. Policies and priorities are influenced by historical developments, climate, geography, economics, culture and the administrative and political environment.

To reflect public health concerns within the proliferation of voluntary green building programs, for example, the US Environmental Protection Agency is developing retrofit guidelines and protocols to be used with energy efficiency programs. These guidelines are intended to prevent possible health harms from energy conservation and to take advantage of opportunities for improvements to be made during conservation efforts.[41]

Building codes gradually are undergoing a review for "green" features. European countries require an Energy Performance Certificate on the sale or rental of a dwelling. In Switzerland the *Minergie* building code is part of the formal regulatory system. Structures built to Minergie requirements are entitled to tax benefits and tend to sell at higher market values.

However, even the most progressive housing codes do not guarantee safe and healthy housing. Many local code enforcement agencies rely on complaints to trigger inspections because of limited resources, or lack sufficient enforcement power to order prompt remediation and impose stringent penalties. Tenants are often reluctant to file complaints for fear of owner retaliation. Thus systematic code enforcement is an important supplement to complaint-based enforcement.

In many developing countries, building codes relate very minimally to issues of insulation and siting and are inadequately enforced. This leaves large populations vulnerable to extreme heat and cold, rain, monsoons and snow, not to mention building collapse in earthquakes, floods and other natural disasters.

A key issue in developing countries is the rapid growth of unregulated peri-urban slums and informal urban dwellings, where building codes are rarely enforced, leaving residents without safe shelter, adequate utility infrastructure, or access to public transport services. A times, however, building codes may be enforced in a selective manner to promote "slum clearance" of unwanted areas or to limit population growth of a minority group.

On the other hand, building policies and codes that merely sanction any kind of urban sprawl embed tremendous energy inefficiencies into future transport, water and sanitation and energy provision. These can have a larger carbon footprint due to their geographical dispersion, implying greater costs for government authorities, utilities and users, as well as possibly higher risk of outages, shortages and, in the case of water supply, system leaks.[42]

Horizontal growth is hardly unique to developing countries and emerging economies. It has been the pattern in many parts of the developed world for half a century or more. In North America, perhaps the major barrier to more energy-efficient housing and urban design has been the ubiquitous development over the past 50 years of single-family-zoned residential neighbourhoods around most towns and cities with large lot provisions. These communities explicitly or implicitly bar mixed-use commercial activity within their boundaries, obligating car travel for even the most basic services such as health clinics, child care and workplaces. Single-use zoning is widely acknowledged to have been a driving force in the development of energy-intensive housing and urban forms reliant on private car transport; this in turn makes North American cities the heaviest transport-related energy consumers in the world per capita (Fig. 8).[43]

Increasingly, European planning policies also relate indirectly to greenhouse gas emissions in terms of requirements for public transport access and walking/cycling systems as well as urban, town and village "densification." For instance, Swiss local authorities may allow property owners to build one- to three attached or semi-attached units in primarily single-family residential neighborhoods. This can create a powerful market incentive to develop more compact and energy-efficient housing in existing villages and towns without sacrificing key health or quality of life features, insofar as duplexes or triplexes are typically more energy-efficient than single-family units, in terms of building

> A key issue in developing countries is the rapid growth of unregulated peri-urban slums and informal urban dwellings, where building codes are rarely enforced, leaving residents without safe shelter, adequate utility infrastructure or access to public transport services.

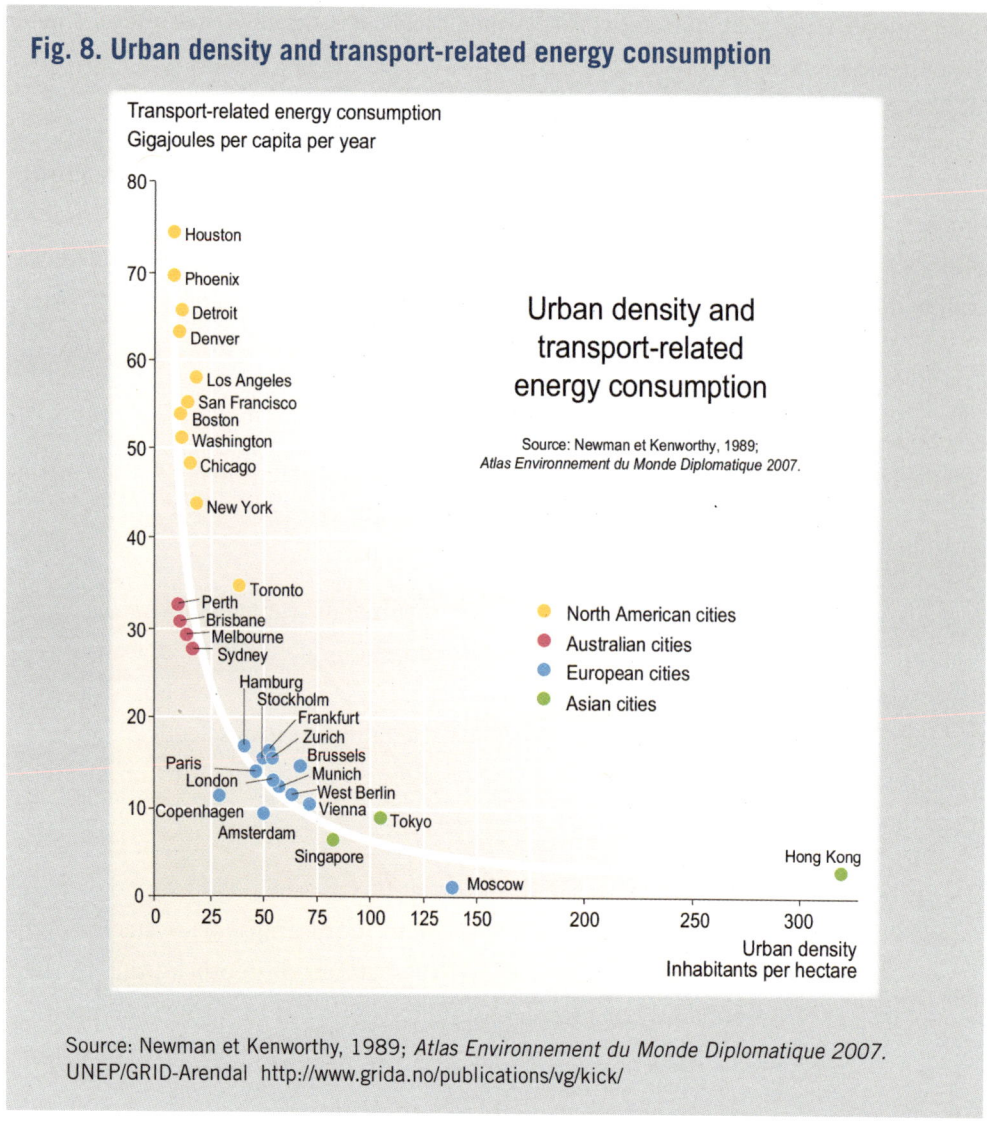

Fig. 8. Urban density and transport-related energy consumption

Source: Newman et Kenworthy, 1989; *Atlas Environnement du Monde Diplomatique 2007*. UNEP/GRID-Arendal http://www.grida.no/publications/vg/kick/

energy use and public transport/access features. Such policies would be unimaginable in many US suburbs. At the same time, strict planning limits on housing development in agricultural and open areas in parts of Europe, e.g. Switzerland, also may be associated with higher overall housing costs, and this in turn can be associated with housing scarcity, landlord abuse of tenants, cross-border housing migration, informal squatting and aspects of homelessness.

Regulations often apply to new dwellings, but in some cases they can have a retroactive effect. Even non-retroactive regulations can in some cases also apply to existing buildings when they undergo important repairs or complete rehabilitation. The interpretation of the terminology "existing" or "new" is not identical in all countries. For England in particular, as soon as a dwelling is completed it becomes an "existing" dwelling, while in Portugal "new buildings" are defined as those built after 1951.

5.5 Tools for financing interventions

Most of the avaialable tools and fiscal instruments used to finance more climate-friendly housing lack explicit consideration of health co-benefits that may, nonetheless, often be derived, as per the discussion in Chapter 3, from improved thermal conditions and overall home energy efficiencies; reduced mould and dampness; and less indoor air pollution from inefficient heating systems, etc.

Available financial tools for financing low-carbon and climate-friendly housing are well summarized in a recent analysis of Green Economy development opportunities by the United Nations Environment Programme (UNEP)[ii]. The analysis, *Towards a green economy: pathways to sustainable development and poverty eradication*, discusses tools in the following categories, to which only a very brief mention is made here: [44]

- tradeable quotas (e.g., cap-and-trade)
- energy performance contracting
- cooperative procurement
- Clean Development Mechanism (CDM) and other flexible mechanisms supporting implementation of the United Nations Framework Convention on Climate Change (UNFCCC) Kyoto Protocol
- regional, national and local tax instruments/incentives for energy efficiency certification schemes.

The Clean Development Mechanism (CDM) is regarded as one of the most important international mechanisms for financing emissions reductions and supporting sustainable development in developing countries. It awards certified "emissions reductions credits" to developed countries investing in projects that reduce emissions in developing countries beyond what would have been otherwise attainable.

However, among the more than 4,500 CDM projects that had been submitted for CDM review (as of April 2009), only 14 addressed energy efficiencies in buildings.[44]

This is largely because the current array of CDM requirements make it difficult or unprofitable for housing projects to demonstrate energy efficiencies that qualify for support. Key barriers, as described by UNEP and other analyses, include: 1) absence of realistic and well-defined building/energy performance baselines against which proposed CDM projects can demonstrate improvements 2) requirements for detailed "technology-focused" demonstration of energy savings, rather than a "whole building" approach that might include a number of integrated measures (e.g. passive design, heating, and lighting). This adds greatly to the administrative cost of a CDM application.

At national and regional level, tradable carbon and energy "certificates" are among the most common market-based policy instruments. Sometimes called rainbow certificates, these include "white" certificates for energy savings, "green" for renewable energy, and "black" for greenhouse gas reductions, and are being implemented through national and regional legislative mechanisms of different countries and regions:[45]

- "Black certificates" are emissions allowances for defined quantities (in tons) of CO_2 over a given period, as managed under the European Union Emission Trading Scheme (EUETS).
- "White certificates" are based on mandatory energy saving targets that certain actors (e.g. power companies and gas suppliers) have to meet by promoting energy efficiency programmes to their customers (including households). They are currently

---

[ii] The UNEP-led *Green Economy* Initiative, launched in late 2008, consists of several components whose collective overall objective is to provide analysis and policy support for investing in green sectors and in more sustainable development in major economic sectors, such as building and transport.

implemented in a limited number of countries, e.g. United Kingdom, Italy, and parts of Australia.
- "Green certificates" may be awarded to producers of renewable electricity for each unit produced, and be traded on a Tradable Green Certificate (TGC) market. This market is driven by mandatory targets for renewable energy production, supply or consumption as set by different (primarily European) countries or others.
- Voluntary certificates may be supported by regional, national or local fiscal instruments or incentives, for example the Swiss *Minergie* voluntary label for energy-efficient housing.

There are also a range of other fiscal instruments, including tax exemptions and loan incentives offered by national, regional and international financial institutions and development banks. Examples cited by the UNEP *Green Economy* analysis include:

- tax exemptions/reductions
- public benefit charges
- capital subsidies, grants,
- subsidized loans or mortgages at lower interest rates
- fee waivers and expedited building applications
- performance-based contracting.

This is only a brief summary of the various financing measures available for climate change mitigation in the housing sector, as well as barriers to their effective use.

While more detailed analysis is beyond the purview of this report, it appears clear that many of the same financial tools that support more climate-friendly and energy-efficient housing have the potential to promote health goals in housing overall. However, it is important that these tools also facilitate access by low-income groups to affordable housing and energy efficiencies.

Notably, serious barriers exist to the wider use of the best-known tool, CDM finance, for housing initiatives, particularly in developing cities where they could be harnessed to shape more sustainable patterns of future growth. These barriers require careful review in the context of global policies on mitigation and mitigation finance.

As part of any review, what is needed is more explicit analysis of *housing and health* co-benefits that may be derived in through the extension of certain kinds of carbon credits and financial support. Incorporating explicit recognition of health gains (and where relevant risks) into the system of international carbon credits, as well as market-based trading and lending would likely open up more opportunities for broader participation in mitigation initiatives that lead to more energy-efficient and healthier housing.

References

1. *A human health perspective on climate change.* Environmental Health Perspectives & National Institute of Environmental Health Sciences, 2010.
2. Ormandy D. *Review of some programmes and initiatives to improve housing and the built environment: background paper for the International Workshop on Housing, Health and Climate Change.* Geneva, World Health Organization, 2010.
3. China National Engineering Research Centre for Human Settlements. *Technical specification for construction of healthy housing.* Beijing, China Planning Press, 2009.
4. Minergie label (http://www.minergie.com/label.html)
5. National Australian Built Environment Rating System (http://www.nabers.com.au/faqs.aspx)
6. Green Building Council of Australia (http://www.gbca.org.au/)
7. Vanzolini (http://www.vanzolini.org.br/)
8. Green Building Council Brasil (http://www.gbcbrasil.org.br/pt/)
9. Canada Green Building Council (http://www.cagbc.org/)
10. Green Globes (http://www.greenglobes.com/)
11. China Green Building Network (http://www.cngbn.com/)
12. VTT Business From Technology (http://www.vtt.fi/)
13. Certivea (http://www.certivea.fr/)
14. German Sustainable Building Council (http://www.dgnb.de/_de/)
15. CEPHEUS – Cost-Efficient Passive Houses as European Standards (http://www.cepheus.de/)
16. BEAM Society (http://www.hk-beam.org.hk/)
17. Indian Green Building Council (http://www.igbc.in/)
18. Green Rating for Integrated Habitat Assessment (http://www.grihaindia.org/)
19. *Green building standards worldwide.* Israel Ministry of Environmental Protection. (http://www.sviva.gov.il/Enviroment/bin/en.jsp?enPage=e_BlankPage&enDisplay=view&enDispWhat=Object&enDispWho=Articals%5El4607&enZone=israel_green)
20. ITACA (http://www.itaca.org/)
21. Green Building Council Italia (http://www.gbcitalia.org/)
22. Comprehensive Assessment System for Built Environment Energy (http://www.ibec.or.jp/CASBEE/english/overviewE.htm)
23. Green Building Index (http://www.greenbuildingindex.org/)
24. Consejo Mexicano de Edificación Sustentable (http://www.mexicogbc.org/)
25. Dutch Green Building Council (http://www.dgbc.nl/)
26. New Zealand Green Building Council (http://www.nzgbc.org.nz/main/)
27. Philippine Green Building Council (http://philgbc.org/)
28. Lidera (http://www.lidera.info/)
29. Building and Construction Authority (http://www.bca.gov.sg/GreenMark/green_mark_buildings.html)
30. Green Building Council South Africa (http://www.gbcsa.org.za/)
31. Minergie (http://www.minergie.com/home_en.html)
32. U.S. Green Building Council (http://www.usgbc.org/DisplayPage.aspx?CategoryID=19)
33. International Living Building Institute (http://ilbi.org/)
34. Green Globes (http://www.greenglobes.com/)
35. Build It Green (http://www.builditgreen.org/)
36. National Green Building Program (http://www.nahbgreen.org/)
37. BREEAM: Environmental assessment method for buildings around the world (http://www.breeam.org/)
38. Abu Dhabi Urban Planning Council (http://www.estidama.org/)
39. Keall M, Baker MG, Howden-Chapman P et al. Assessing health-related aspects of housing quality. *Journal of Epidemiology and Community Health,* 2010, 64(9):765–71. doi:10.1136/jech.2009.100701
40. *Housing health and safety rating system: operating guidance.* London, Office of the Deputy Prime Minister, 2006.
41. *Healthy indoor environment protocols for home energy upgrade.* Washington, US Environmental Protection Agency. (http://www.epa.gov/iaq/homes/retrofits.html)
42. Jordan, water is life. In: *Health and environment: managing the linkages for sustainable development, a toolkit for decision-makers.* Geneva, World Health Organization, 2008.
43. Kenworthy J et al. *Transport, health and developing cities policy brief.* Geneva, World Health Organization, 2009. (http://www.who.int/heli/risks/urban/urbanenv/en/index.html, www.who.int/heli/)
44. *Towards a green economy: pathways to sustainable development and poverty eradication.* Nairobi, United Nations Environment Programme, 2011.
45. Bonnville E, Rialhe A. *White, green and black certificates: three interacting sustainable energy instruments.* In: *Energy Policy,* October 2005. Accessed at: www.leonardo-energy.org.

6

India: Children study at night under the glow of electric lights charged by solar-powered PV stations, established in villages under the "Lighting a Billion Lives" initiative.

The Energy and Resources Institute/TERI

Case studies of good practice

This section focuses on examples of effective housing interventions that reduce energy consumption and CO_2 emissions and result in improved health for inhabitants. Four case studies address: a) retrofitting houses with insulation and resulting health, energy use and CO_2 savings; b) use of solar and photovoltaic lighting in small households in developing countries; c) a low-cost urban housing energy upgrade and d) habitation improvements to reduce vector populations.

6.1 New Zealand's *Housing, insulation and health* study

The *Housing, insulation and health* study is a cluster randomized trial of retrofitting insulation in predominantly low-income communities in New Zealand. The study was carried out in 2001–2002 to assess whether installing insulation in houses affects occupants' health as well giving energy and environmental co-benefits.[1,2]

The interventions included ceiling insulation, draught-stopping around windows and doors and fitted insulated paper installed beneath floor joints, as well as a polythene moisture barrier on the ground beneath the house. Some 1350 households (4407 people) were randomly selected in which at least one person had symptoms of respiratory disease.

Health data were collected through self-reported measures of health, comfort and well being, primary care (GP) visits, and days off school and work. With regards to energy consumption, complete records were collected over the two-year study life on self-reported use of bottled LPG, wood and coal. However energy savings could only be reliably quantified on the basis of metered data such as electricity and gas usage. On the basis of metered energy savings, greenhouse gas emission reductions were estimated.

6.1.1 Mitigation impact

A "typical" household – one with a heating pattern typical of the weighted average of all households in the study for which there was good data – benefited from net energy savings (metered electricity and gas) of 13%. In absolute terms, intervention group energy use declined 5% while control group use increased 8%.

This net saving amounts to around 532 kWh over a year. Although the reduction in metered energy use was not statistically significant at 5% given the limited number of households for which both years of records were available, the reduction in energy use based on both measured and self-reported energy use was statistically significant at the 1% level (p=0.0006). Because it was more difficult to price non-metered energy sources such as firewood and coal, these were not considered in the energy savings calculations, although self-reported decline in use could enhance this mitigation benefit. A subsample of electricity consumption in 116 houses in one city showed a significant average decrease in peak period demand of 25.5%, from 2.15 kW to 1.60 kW. This has significant regional and potentially national implications for power generation, as it is peak demand that drives the need for electricity generation, transmission and distribution capacity at the margin.[1]

6.1.2 Health impact

Improved insulation was associated with a small average increase in bedroom temperatures during the winter (0.5° C) and decreased relative humidity (−2.3%). Bedroom temperatures were below 10° C for 1.7 fewer hours each day in insulated homes compared with uninsulated ones.

Changes in temperature were associated with reduced odds in the insulated homes of fair or poor self-rated health (adjusted odds ratio 0.50, 95% confidence interval 0.38 to 0.68), self-reports of wheezing in the past three months (0.57, 0.47 to 0.70), self-reports of children taking a day off school (0.49, 0.31 to 0.80), and self-reports of adults taking a day off work (0.62, 0.46 to 0.83). Visits to general practitioners were less often reported by occupants of insulated homes (0.73, 0.62 to 0.87). Hospital admissions for respiratory conditions were also reduced (0.53, 0.22 to 1.29) but this reduction was not statistically significant (P=0.16).

6.1.3 Cost-benefit

The results of the study suggest that total benefits in "present value" terms are 1.5 to 2 times the magnitude of the costs of retrofitting insulation, with the health benefits being relatively greater in terms of present-day economic valuation.

6.1.4 Conclusion

Retrofitting dwellings with insulation and thermal envelope improvements provides health benefits as well as energy savings. The relatively large present-day economic value of stream of health benefits, particularly in terms of reduced hospital admissions, provides good justification for energy efficiency schemes even when CO_2 savings, are more difficult to measure and represent relatively less economic value in present-day economic terms.

6.2 Lighting a Billion Lives

Around 1.5 billion people in the world lack access to electricity; about a quarter of these live in India. Forced to light their homes after sunset with kerosene lamps, dung cakes, firewood and crop residue, millions live with constant risks to their health. Along with impacts of indoor air pollution already described, use of kerosene for lighting has been

Fig. 9. Solar lanterns – lighting up homes in India

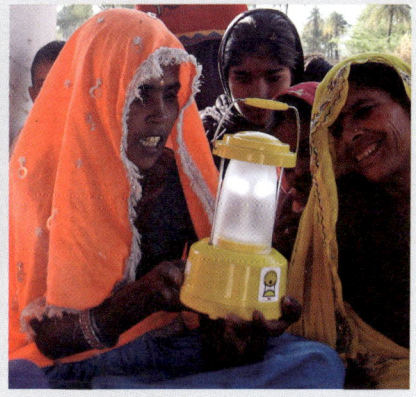

Photo: The Energy and Resources Institute (TERI)   Photo: The Energy and Resources Institute (TERI)   Photo: Ashden Awards for Sustainable Energy (www.ashdenawards.org)

associated in some studies with increased TB incidence, illustrating a health issue specifically related to household lighting. There is also a risk of injury from tipping over kerosene and paraffin lanterns.

A large Indian NGO, the Energy and Resources Institute (TERI), has set out to address these issues in a major government-backed campaign. The "Lighting a Billion Lives" (LaBL) initiative is providing thousands of rural Indian households with ultra-low-energy LED lanterns powered by sunlight. Recharged daily by a PV solar panel, a lantern can provide 3–6 hours of light to a household each night.[3] The initiative was launched in 2007 at the Clinton Global Initiative, and within three years the solar lanterns had spread rapidly, reaching 600 villages in 16 Indian states and providing light for 160,000 people. TERI has established solar photovoltaic charging stations in each village to recharge the lanterns; these are operated by local entrepreneurs who are selected and trained by TERI and other grassroots partners. The lanterns (Fig. 9) are rented daily to households and enterprises for a small fee (less than a quarter US dollar). The TERI initiative offers seven solar lantern models with varying degrees of power.

6.2.1 Mitigation impact

The initiative facilitates socioeconomic development of the village while offering local and global environmental benefits. Each solar lantern displaces the use of about 40–60 litres of kerosene a year and an estimated 400–500 litres of kerosene in a lifespan of 10 years, thereby mitigating about 1.45[iii] tonnes of CO_2.

6.2.2 Health impact

Users, particularly women, reported a significant reduction in frequent cough and eye rashes as a result of replacing kerosene or paraffin lamps with the solar lanterns. They also report fewer accidents and injuries due to toppling of kerosene lamps. At the household level, the initiative helps reduce smoke pollution in rural women's indoor work environment, resulting in fewer complaints of red eyes and heavy headaches.

---

iii At emission intensity of 2.45kg CO_2/litre of kerosene

In several villages, the solar lamps are being used by midwives for safer delivery of babies. For instance, in Dhulkot village of Madhya Pradesh, the solar lanterns are being used in the primary health centre, which has a more complex solar-powered electricity system with AC current dependent on an inverter. When the inverter is out of order, electricity is not available from 6 pm to 10:30 pm. The solar lantern illustrates the power of use and development of small-scale DC-powered solar devices, which are easier to operate, replace or repair.

6.1.3 Socioeconomic impacts

The lanterns have rapidly found a central place in rural households, health centres and culture. Villagers describe the lights as the "saviour in the darkness that otherwise envelops the kitchen." Light is now available not only for cooking but also for animal-tending chores, for children's studies, for midwives and for safety when walking at night or guarding village perimeters.

The entrepreneur model used for the charging stations illustrates the sustainable development co-benefits of such small-scale solar development. The solar lanterns also allow extended working hours for rural communities in fields and shops, longer studying hours for children and safer movement for village elders at night. In many villages, the initiative has facilitated small-scale industry activities, such as betel-leaf farming in West Bengal, eco-tourism activities in the tribal areas of Orissa and basket-making cottage industry in Rajasthan.

6.3 Low-cost urban housing energy upgrade project in Cape Town, South Africa

> The Kuyasa project is is the first activity in South Africa, and one of a handful of housing initiatives globally registered under the Clean Development Mechanism. It demonstrates how climate finance could be harnessed more widely to improve health and the environment.

This is the first activity in South Africa, and one of a handful of housing initiatives globally, that has qualified for registration under the UNFCCC, Clean Development Mechanism (CDM). It demonstrates how climate finance could be harnessed more widely to improve health and housing environments in low-income countries.

The Kuyasa project aims to improve the thermal performance and reduce CO_2 emissions of both existing and future housing units in Kuyasa, a low-income neighborhood of Khayelitsha Township in southeastern Cape Town. Retrofits of exisiting housing have involved installation of solar water heaters, ceiling insulation and low-energy, long-life compact fluorescent light bulbs (CFLs). Since its launch in 2008, the programme has involved upgrades in about 2300 units, and is planned for a period of 21 years. The project is being carried out by the City of Cape Town in collaboration with the *South-SouthNorth Project,* a network of institutions and experts that helps public and private stakeholders in Africa and South-East Asia navigate the CDM qualifying process.

6.3.1 Mitigation impact

The project is projected to save about 2.8 tons of CO_2 per household, per year, at a savings of about US$ 110 per household annually in energy costs.[4]

A typical Kuyasa interior, prior to retrofit. The exposed tin roof radiates heat in the summer and absorbs cold and moisture in the winter. Ceiling insulation being added as part of the improvements should help reduce damp and temperature extremes, and save energy. (Photo: Nic Bothma, Kuyasa/CDM Project)

6.3.2 Health impact

The upgrading measures implemented through the Kuyasa project have reduced the rate of condensation and dampness in the dwellings and decreased the risk of disease, especially tuberculosis. As also noted in Section 3, increased access to hot water has been linked to improved hygiene through reductions in the bacterial load as a result of washing. This can in turn reduce diarrhoeal diseases. Other co-benefits of the project include reduced local air pollution and hence reducing pulmonary pneumonia, carbon monoxide poisoning and other respiratory illness, as well a less fuel and energy poverty. The lack of proper ceilings and insulation meant that families were spending around 10 rand a night (about US$ 1.5) to heat their homes in winter.

6.3.3 Cost-benefit

Residents are billed a small amount each month for the improvements, but this is offset against the direct savings in monthly heating bills as a result of the project. Labour for the installation is sourced in the community and each head of household is also given a small stipend to facilitate installation and maintenance. The technology used is local. While imported stainless steel water tanks may be cheaper, the project team at Kuyasa realized that the high levels of chlorine in local water (used to kill bacteria) corrodes stainless steel when water is heated. The project team opted instead for a locally made product. This has not only kept the revenue in the country, but the longevity of the local product has proved to be a long-term cost saving.

6.4 Low-carbon housing measures and vector-borne disease control

6.4.1 Home improvements to reduce Chagas disease

Housing improvements have long been used as a preventive strategy against Chagas disease in southern Latin America. The disease is spread by triatomine bugs that traditionally lived primarily in forests, but over the past century have become a much more serious problem in rural and urban dwellings as a result of urbanization, deforestation and increased rural/peri-urban development. Housing improvements shown to reduce triatomine vector infestation include concrete floors, plaster, brick walls and tiled roofs.[5] A more recent study in Jutiapa, Guatemala, evaluated seventeen variables as possible risk factors for infestation with *Triatoma dimidiata*. During 2004, 644 houses were assessed for vector presence and evaluated for hygiene, cluttering, material comfort, construction conditions and number of inhabitants, among other factors. The study showed a greater chance of vector presence when walls lacked plaster (3.85 times) or had low-quality incomplete plastering (4.56 times) compared with walls that were completely plastered, as well in houses with poor sanitation and other construction conditions.[6]

6.4.2 Dengue control through water storage container management

The most productive larval habitats for *Aedes* mosquito reproduction typically are water storage containers in and around houses where piped water is unavailable, and debris (e.g. old tires) which accumulate rainwater and thus provide breeding grounds. Systematic reviews and field studies have identified methods to identify the most "productive" larval habitats and then target these habitats with vector control interventions. These control measures can include many environmentally friendly interventions such as water storage, sanitation and waste cleanup measures that can in turn reduce reliance on chemical spraying. Covering water containers and larviciding water containers, including with biological predators or agents, have been found to be particularly effective, dependent on the local eco-epidemiological setting.[7-12] The WHO/UNDP/UNICEF/World Bank Special Programme for Research and Training in Tropical Diseases (WHO/TDR) is carrying out a series of field studies on improved eco-bio-social management of dengue and Chagas in nine Latin American and six South-East Asian communities in order to fine-tune intervention strategies based on good environmental management.

6.4.3 Lime-plastering of walls in visceral leishmaniasis control

Typically, sand fly vector control in India is carried out with indoor residual spraying (IRS), often with DDT, a use of persistent organic pollutants (POPs) that is permitted under the Stockholm Convention on vector borne-disease control.

One recent intervention study carried out under the auspices of WHO/TDR found that traditional lime-plastering of walls, while not quite as effective as DDT, still had excellent efficacy in reducing sand fly vector densities over a five-month period in two Indian and Nepalese study sites.[13] Further exploration of low-carbon housing improvements that promote visceral leishmaniasis vector control, in conjunction with long-lasting insecticide-treated nets or bednets, is warranted particularly in light of other potential health co-benefits of housing improvements as well as the carbon and environmental footprint of DDT.

6.4.4 Overall mitigation impacts

All studies mentioned above analysing use of housing interventions like lime-plastering and better house construction features reduce the use of DDT and other chemical vector control tools that have a carbon footprint. As households develop socioeconomically, effective low-carbon vector control methods including bednets and screens can also help reduce the tendency to shift to air conditioning to prevent insect biting, particularly at night.

6.4.5 Overall health impact

Chagas, visceral leishmaniasis and dengue have serious morbidity and mortality, and with dengue representing the world's fastest-growing arbo-viral infection, reducing the incidence of these diseases is a major public health priority. At the same time, the use of chemical pesticides, particularly DDT, in vector control is associated with a range of immediate and chronic health impacts. "Judicious use" is a key principle of vector control, as described in the WHO Global Strategy for Integrated Vector Management (IVM). Better housing and household water management can be a key means of implementing IVM strategies into other sustainable development initiatives. Such control strategies should not only reduce or eliminate infestation, but also prevent vector reinfestation.

References

1. Chapman R et al. Retrofitting houses with insulation: a cost-benefit analysis of a randomised community trial. *Journal of Epidemiology and Community Health*, 2009, 63(4):271–277.

2. Howden-Chapman P et al. Effect of insulating existing houses on health inequality: cluster randomised study in the community. *British Medical Journal*, 2007, 334:460.

3. *Lighting a billion lives* [brochure]. New Delhi, The Energy and Resources Institute. (http://labl.teriin.org/pdf/labl-Brochure-low%20Res.pdf).

4. *Kuyasa Energy Efficiency Project*. Environmental Resource Management, City of Cape Town. Accessed at: (http://www.capetown.gov.za/en/EnvironmentalResourceManagement/projects/ClimateChange/Pages/KuyasaEnergyEfficiencyProject.aspx).

5. De Andrade et al. Evaluation of risk factors for house infestation by *Triatoma infestans* in Brazil. *American Journal of Tropical Medical Hygiene*. 1995, 53(5):443–7.

6. Bustamante DM et al. Risk factors for intradomiciliary infestation by the Chagas disease vector *Triatoma dimidiate* in Jutiapa, Guatemala. *Cadernos de Saúde Pública*, 2009, 25(Sup. 1):S83–S92.

7. Focks DA, Chadee DD. Pupal survey: an epidemiologically significant surveillance method for *Aedes aegypti*: an example using data from Trinidad. *American Journal of Tropical Medicine and Hygiene*, 1997, 56(2):159–167.

8. Focks DA. *A review of entomological sampling methods and indicators for dengue vectors*. Geneva, Special Programme for Research and Training in Tropical Diseases / World Health Organization, 2004.

9. Kay B, Nam VS. New strategy against *Aedes aegypti* in Vietnam. *Lancet*, 2005, 365(9459):613–617.

10. Focks DA et al. Transmission thresholds for dengue in terms of *Aedes aegypti* pupae per person with discussion of their utility in source reduction efforts. *American Journal of Tropical Medicine and Hygiene*, 2000, 62(1):11–18.

11. Focks DA et al. The use of spatial analysis in the control and risk assessment of vector-borne diseases. *American Entomologist*, 1999, 45(3):173–183.

12. Arunachalam N et al. Eco-bio-social determinants of dengue vector breeding: a multicountry study in urban and periurban Asia. *Bulletin of the World Health Organization*, 2010, 88(3):173–184.

13. Joshi AB et al. Chemical and environmental vector control as a contribution to the elimination of visceral leishmaniasis on the Indian subcontinent: cluster randomized controlled trials in Bangladesh, India and Nepal. *BMC Medicine*, 2009, 7:54.

7

Nic Bothma/Kuyasa CDM

Kuyasa, South Africa: Workers install new sewage pipes in this Cape Town neighborhood housing and energy efficiency upgrade, a climate change migitation initiative.

Conclusions and recommendations

7.1 Largest health co-benefit opportunities

This review examined the health co-benefits and risks from IPCC-reviewed strategies to mitigate climate change in the housing sector. In the IPCC assessment, housing is regarded as a sector with the greatest immediate potential for cost-effective mitigation of climate change.

This review finds potential large health co-benefits from many of the same measures in terms of opportunities to reduce housing-related chronic diseases, including asthma and allergies, and also chronic lung disease in poor developing country households dependent on biomass and coal energy. Prevention of key airborne diseases such as TB, and also water and sanitation-related diseases, may be achieved through many of the same strategies. The review also identifies health risks that may be created by mitigation measures and how those may be avoided.

Even without reference to carbon gains, the immediate health co-benefits of certain mitigation strategies (e.g. thermal envelope improvements) may more than justify new investments in terms of the costs of avoided sick days, doctor visits and hospitalization. Considering health savings along with energy savings, makes for a better evaluation of costs and benefits to society of housing interventions motivated by climate change. Health economics can therefore be an important driver of mitigation economics. By promoting more effective housing and health strategies that avert significant disease, the health sector also can make better use of its own resources. This is a critical message for health systems worldwide.

Specific recommendations are presented here, both in terms of policy measures for which sufficient evidence now exists to take action, and issues requiring further research/evaluation to develop and fine-tune policies.

7.1.1 Optimize health benefits of insulation and home heating retrofits

- Improved "thermal envelope" and heating efficiencies (e.g. replacing open-flame gas and electric space heaters, or inefficient coal/biomass stoves, with cleaner alternatives) can significantly reduce both acute and chronic respiratory illness, including asthmas and allergies. It is essential to ensure good ventilation to reduce build-up of harmful indoor air pollutants, as well as adequate daylight to prevent mould growth and support good mental health.

- Policy-makers, planners and regulators should work to ensure that heating, cooling and thermal envelope innovations are affordable and accessible to poor households.

7.1.2 Emphasize active and passive natural ventilation in cooling strategies

- Natural ventilation is a key health parameter of indoor air quality. Use of active and passive natural ventilation with appropriate humidification or dehumidification measures can help prevent airborne infection transmission and reduce other chronic respiratory conditions.
- In malaria-endemic countries, natural ventilation should be combined with reinvigorated policies for use of window/door screens and bednets.
- In heavily polluted urban areas, natural ventilation measures would have to be accompanied by appropriate mechanical filtering of air, at least until ambient urban air quality improves.

7.1.3 Make use of environmental cooling measures that also promote healthier neighbourhoods

- Tree planting, green spaces, ponds or water fountains, and courtyards are all design measures that support natural cooling. These same measures are health-enhancing insofar as they can support better thermal conditions and more effective use of natural ventilation, as well as mental health, and, in certain settings, hygiene/sanitation. Such environmental cooling features are often embedded in traditional urban and building design principles of many cultures, and deserve renewed emphasis.

Drawing water from a spring in a Nepalese village. (Photo: Heather Adair-Rohani)

7.1.4 Integrate climate change adaptation and mitigation

Climate change adaptation measures undertaken today for housing should not contribute to greater future climate change impacts that harm health later (via postponed risks).

Mitigation measures should consider the need to adapt to immediate climate-related threats, such as heat waves, flooding and extreme weather. A new "adaptive mitigation" paradigm is proposed to integrate the best knowledge from both adaptation and mitigation arenas with respect to health, and to obtain the best overall package for health, cost-effectiveness and long-term environmental sustainability. This can include features such as:

- "Resilient and climate-friendly" housing designs in cities and areas vulnerable to climate-related heat waves, disasters and/or other types of natural hazards (e.g. wind, landslide, flood, earthquakes).
- Climate-adapted shelter that protects against environmental risks from dust, insects and rodents, unsafe water and sanitation, noise, crime and violence.
- Cleaner biomass and biogas household stoves in developing countries that reduce indoor air pollution and climate-related fuel poverty as well as climate change.
- Passive solar-heated homes that provide energy resilience in cold climates, while reducing energy emissions for heating.
- Passive solar-powered hot water, solar photovoltaic electricity and low-energy DC solar lights and other appliances that provide energy resilience for the poor in developing countries, as well as improving health (e.g. respiratory diseases, hygiene, food safety, eyesight, well-being).

> Mitigation measures should also improve resilience to immediate threats from climate change, such as extreme weather. Adaptation measures should contribute to reducing climate change emissions in housing over the long-term.

7.2 Health risks to be avoided

To achieve mitigation, well-known measures to avoid health risks should systematically be integrated as part of the implementation. These include: 1) provision for adequate ventilation in the case of thermal envelope improvements and active or passive cooling systems, 2) protection from vector borne disease in the case of naturally ventilated buildings in malaria-endemic climates; and 3) use of healthy and safe building materials, including avoidance of toxics and carcinogens, such as asbestos and lead pipes and paints.

Since a large portion of carbon savings by 2030 will be achieved through the retrofit of existing buildings and replacement of energy-intensive appliances with more efficient ones, it is essential that health-relevant policies address building use behaviours as well as health at various stages in the building life cycle, from construction to retrofit and demolition, when risks of exposure to harmful materials such as asbestos may be particularly acute.

7.2.1 Ensure adequate ventilation

Measures improving the thermal envelope should ensure adequate ventilation in order to reduce risks of indoor air pollution and enough windows for day lighting.

7.2.2 Avoid harmful materials; treat others with care

Improved "thermal envelope" measures should avoid use of health-harming materials. Based on available evidence, the following materials should be avoided in building construction, insulation and repair activities: asbestos, lead paint, pressed wood products manufactured with volatile organic compounds (e.g. formaldehyde), arsenic in timber, batt insulation materials containing formaldehydes, and foam boards containing carcinogens and endocrine disruptors. Healthy housing initiatives can support greater awareness of harmful materials in the contexts of national policies and codes, consumer awareness and occupational practices.

Further study also is needed on which insulation materials should be recommended to replace dangerous materials, particularly asbestos in developing countries.

7.3 Gaps in current mitigation analysis

7.3.1 Optimize housing mitigation and health co-benefits through better urban planning

There is much evidence that housing in more compact urban forms and in mixed residential/commercial uses can enhance many health-relevant aspects of the residential environment, including: more travel by walking, cycling and transit and thus reduced air and noise pollution from vehicle travel; more space and opportunities for safe physical activity; greater independent mobility for children, elderly and other vulnerable groups moving in and around the neighborhood.

These very key aspects of land use planning, critical to both health and climate change mitigation, are not addressed in detail by the IPCC Fourth Assessment report. It is hoped that housing mitigation in the urban context will be more fully addressed in the forthcoming Fifth Assessment report.

The "whole building" approach towards energy efficiency of housing structures should be extended to a "whole neighbourhood" approach to both mitigation and health, identifying co-benefits at community level of more energy-efficient urban housing design.

Similarly, there needs to be more systematic assessment by the health sector of the health gains and tradeoffs of different forms of urban design (high, medium and low densities), and at different levels (household, neighbourhood and community).

Although urban planning and land-use changes are often perceived by health policy-makers as costly and difficult to implement, case studies from around the world show that these changes are not only happening in many settings, but also can often be among the most cost-effective means of building lower-carbon housing and cities in the long term.

> **Avoid harmful materials like asbestos, lead in paint, arsenic impregnated timber products and pressed wood products with formaldehyde binders. Construction workers are especially exposed to harmful building products, and should be protected.**

7.3.2 Emphasize healthy and sustainable development for rapidly growing low-income cities

As noted by IPCC, the largest savings in future energy use (75% or higher) can be realized through the better design of new buildings as complete systems. Here, too, great health gains may be obtained in terms of improved possibilities for natural ventilation, reduced exposure to heat stress during heat waves, better thermal conditions, and more optimal use of renewable energy sources to reduce air pollution exposures.

It is in low- and middle-income cities, where urban growth is very rapid, that the greatest opportunities exist for a "systems" approach. As a result, this may be where some of the greatest health co-benefits from mitigation measures can be derived, and future climate change impacts avoided.

Yet low-income housing settings have not been, so far, a strong focus of mitigation analysis as reflected in IPCC review.

Health co-benefits are realized when housing improvements enhance energy efficiencies, improve access to cleaner and renewable household energy, solar-heated hot water, PV lighting and appliances, and climate-resistant housing structures. Some examples of these have been provide in the Chapter 6 review of the TERI "Lighting a Billion Lives" housing initiative in India and the Kuyasa, Cape Town, South Africa, Clean Development Mechanism housing improvement scheme.

The gains are even larger when housing is integrated into broader strategies for better urban design and transport, safe drinking-water and sanitation provision, and safe siting.

- "Pre-emptive" development of water/sanitation and public transport infrastructure in new areas, particularly on urban peripheries, will expand access to health-promoting urban services more equitably, thus improving health outcomes, while also reducing the carbon footprint and costs of urban expansion.

Bus Rapid Transit (BRT) is a key feature in a number of major Latin American cities. Portrayed here is Santiago, Chile. (Photo: SpecialistStock)

- Housing and neighbourhood densities and design features such as building heights should facilitate safe, independent mobility of and access for children, older adults, women and other vulnerable groups.
- A shift to alternative fuels and renewable energy sources can assist in reducing pollution from transportation and building operations use as well as overall GHG emissions associated with residential development.
- Low-carbon siting of housing including orientation, clustering and inclusion of green spaces can optimize ambient atmospheric conditions in residential neighbourhoods and reduce their "heat island impact." Siting in many cities needs to be far more carefully controlled to avoid areas at severe risks of flooding, landslides, etc., or to take appropriate protective measures.

7.3.3 Address slum housing in the health and mitigation context

Slums have a tremendous immediate health impact, but their potential long-term climate footprint should be appreciated as well. Large blocs of low-density, informal settlements typically make poor use of urban space, exacerbate the urban heat island effect and make infrastructure delivery (e.g. transport, utilities) more energy-intensive and costly.

Low-carbon, low-cost and health co-benefit "packages" that yield immediate benefits to slum areas are therefore important. Such measures also can address a key Millennium Development Goal currently lagging (MDG 7: improve lives and health of urban slum dwellers), and help address geographic and income/health inequities in the context climate change mitigation.

7.3.4 Optimize the health equity benefits of low-energy systems and renewable energy systems for the poor

There appears to be significant health equity benefit from many forms of renewable energy use in poor homes. These strategies and technologies and their potential benefits include the following, and should be studied and evaluated more systematically:

Passive solar "combi" hot water heating systems cover rooftops in a village near Beijing, China. The systems are also used for space heating, replacing traditional wood stoves. This has improved thermal comfort in the winter, raising indoor room temperatures that previously dipped as low as 5–8 °C at night. (Photo: He Jianqing)

- Replacement of kerosene lamps with photovoltaic solar-powered electricity and low-energy LED lights; research is needed to clarify their impacts on respiratory and eye diseases and on social determinants of health.
- Passive solar-powered water heating for poor households and sanitation impacts.
- Access to low-energy solar DC-powered refrigeration for poor households and food safety/security.
- Provision of house screening, structural improvements and natural ventilation that impact TB and vector-borne disease transmission.
- Carbon-neutral or low-carbon biogas/gasified biomass household fuels and their impacts on respiratory, sanitation-related and parasitic diseases (e.g. schistosomiasis).

7.4 Implementing win-win health, housing and climate change strategies

7.4.1 Undertake health-oriented assessment of policies and plans

Health impact assessment (HIA) systematically considers evidence of health impacts expected from specific policies, plans or projects to mitigate climate change in housing. HIA considers both quantiative and qualitative evidence, and engages stakeholder views and perspectives. The health action plan emerging from an HIA can help make a housing intervention positive for health in the short- and the long-term.

Stakeholder engagement helps close the gap in knowledge about what works best in the field. Community participation helps to identify culturally appropriate and acceptable interventions.

HIA is also a process, and a tool, whereby public health policymakers and practitioners can work closely with housing, energy and planning sectors to harmonize health objectives with urban development and climate change mitigation goals.

Results from HIA can be useful to the primary health care system in identifying housing and health risks, and mitigation co-benefits, and in supporting the population and local leaders to adopt appropriate responses.

> Health impact assessment is an important tool for maximizing health gains, and avoiding health risks, from climate change mitigation measures and investments in housing and urban development.

7.4.2 Support linked health and climate research, monitoring and evaluation for better policies

There is a dearth of evidence on how mitigation strategies impact health. Intervention studies can help document actual gains for health and climate change mitigation. In addition, development of linked health and energy efficiency indicators can support monitoring and evaluation of policies.

More health-oriented intervention studies on specific housing mitigation measures are needed to identify key opportunities as well as risks to avoid. While models are important for scenario analysis, health impacts can be confounded by multiple variables, (e.g. impacts of reduced indoor air pollution from buildings may be confounded by smoking). This underlines the importance of real-world studies of actual interventions.

Models estimating the health co-benefits of climate mitigation by linking emissions of key pollutants with dispersion models and epidemiological evidence require fine-tuning. While the approach is valuable, it can be difficult to apply in the case of broad-based energy-efficiency programs, as these programs tend to affect multiple sources simultaneously (e.g. numerous homes within a state, all power plants on a grid). This review also identified the need for research to:

- Quantify health benefits of actual energy efficiency programmes using harmonized methods and approaches; adequate characterization of uncertainties; and control/adjustment for confounders.
- Examine public attitudes towards, and acceptance of, healthy and low-energy housing measures, including implications of increased natural ventilation versus air conditioning (e.g. for vector-borne disease transmission; personal security; equity, etc).
- Identify best strategies for, and health co-benefits of, climate change mitigation relevant to housing in slums and informal settlements.
- Assess optimal ways to combine housing mitigation and adaptation strategies in diverse regions and economies.

7.4.3 Build capacity for better data collection and monitoring systems on housing and health

There is a need to build capacity within local and national authorities' health services, housing agencies and educational programmes for housing and health professionals. Architectural schools and schools of public health should study healthy housing as part of green design and public/environmental health commitments.

7.5 Regulatory frameworks

7.5.1 Include health more systematically in housing codes, standards and enforcement

- Housing codes and standards are critical to ensuring basic structural soundness and safety, as well as other aspects of housing essential to good health, e.g. water, sanitation and utility access, protection from extreme weather, vectors, and access to pedestrian-safe walking routes and public transport.
- Health needs to be systematically included in all aspects of development of these standards and codes as well as a primary factor in their enforcement.
- Housing codes also are important to ensuring safe siting of housing that protects inhabitants from risks such as mudslides, severe weather (e.g. floods, tsunamis, earthquakes, etc).
- Codes also enforce policies and standards regarding use of harmful and hazardous housing materials such as lead and asbestos, and appropriate labeling of materials.
- In many regions, more systematically-designed codes and regulations, as well as enforcement, are needed to ensure the most basic public health parameters of housing.

7.6 Climate finance for health

7.6.1 Improve finance mechanisms and incentives for all regions and income groups

Current requirements of the United Nations Framework on Climate Change (UNFCC) Clean Development Mechanism (CDM) finance are generally too complex for poor countries and communities to use effectively in the housing context, particularly in the case of smaller housing projects. The technology-specific (as compared to performance) orientation of CDM also makes it difficult, and thus expensive, to demonstrate climate gains through a diverse package of housing measures – even when those may be very large in reality.

Climate finance mechanisms also fail to consider the relative health co-benefits of proposed strategies.

Development banks, donors and national level actors often do not appreciate how housing can be a driver for benefit in both health and climate change mitigation.

The role of carbon and development finance in promoting healthy and climate-friendly housing should be examined to see how key actors can optimize potential health gains in the context of housing finance.

In developing countries, donor programmes, carbon finance, and multilateral/bilateral financial incentives should also emphasize, in particular, transition to cleaner household fuels as well as slum improvement in light of their very large health risks and potential benefits.

Climate finance mechanisms have failed to consider the health co-benefits of green housing strategies. The Clean Development Mechanism (CDM) and other climate financial tools should consider health performance criteria in resource allocation.

Slums growth in fast-growing developing cities has major health impacts, and is yet to be addressed systematically in the context of climate change finance. Here, children walk through the alley of a Chinese slum lined with broken water containers. (Photo: © UNEP / Mark Edwards, Hard Rain Picture Library)

7.7 Building community capacity

Community participation in healthy housing and climate change mitigation can incorporate local knowledge into housing and land use policies. Such efforts also build self-reliance and awareness of problems and solutions. Community representatives typically have grassroots connections and enjoy higher local trust than outside experts. They are well positioned to promote efforts that improve public health and environments.

Community development should enable people living in low-income neighbourhoods to gain organizational experience in dealing with government agencies and other institutions. There should be provision for accountability and sharing of experiences about housing improvements that contribute to health and how health can be integrated into housing and climate policy agendas.

7.7.1 Promote healthy housing behaviours to make best use of climate-friendly housing measures

There is a need to raise awareness of how certain measures that waste energy may also harm health, e.g. continuous use of air conditioners in closed spaces without adequate natural ventilation and, on the other hand, of strategies that prove to be energy-efficient and health promoting:

- Bednet and window/door screen use can help prevent transmission of vector-borne disease.
- Exposure to the outdoors and to natural light sources is essential to health, and to vitamin D requirements in particular. Opportunitites for outdoor activity in and around the home are particularly important to women and children, who tend to spend more time close to home and neighbourhood.

Residents of Mexico City's Chalco district plant tree seedlings provided free by the government to alleviate air pollution and urban heat island impacts of the city. (Photo: © Mark Edwards, Hard Rain Picture Library)

- Active travel (walking and cycling) around neighbourhoods for daily routines should be promoted as a means of healthy physical activity, along with adequate neighbourhood provision for safe street crossings, sidewalks, bike lanes, etc.
- In cold climates, activity also helps regulate body heat. During heat waves, adequate hydration is important. Climate-adapted clothing and bedding measures are important in all climates, and can also help reduce household energy demand.
- Shun indoor tobacco use; if occupants smoke, they should do so outside.
- Promote greater consumer awareness in poor countries of very hazardous construction materials, such as lead and asbestos.
- Promote awareness about interior design and consumer products containing health-harmful materials, e.g. carpet glues and pressed-wood products with formaldehyde.
- Use more efficient lighting as a major mitigation opportunity, but ensure safe disposal of mercury-containing CFL lights, and adequate light to avoid injuries.
- Encourage waste reduction, recycling and reuse through community participation.

7.7.2 Foster a 'primary preventive' approach to healthy housing

Greater knowledge about healthy mitigation measures and tools supports self-reliance in primary prevention of housing-related illness. Integration of this knowledge into primary prevention strategies helps poor and vulnerable communities to set priorities, troubleshoot, identify solutions, and look after their own health in the broadest sense of the term.

7.7.3 Support healthy housing environments for children

A healthy home environment is critical to child health since children spend a large amount of time indoors and are exposed to indoor pollutants at a critical life stage. A healthy and safe neighbourhood facilitates a child's physical and mental exploration of his/her space, and testing of motor and intellectual skills. Unsafe environments not only threaten a child's physical health, they also limit a child's developmental opportunities. Among the recommended actions:

- Health and housing policies should consider more systematically the impacts of urban environmental risks on children's health and development, and incorporate that knowledge into actual planning and renovation of housing environments.
- Health Ministries should build capacity among health professionals to recognize and manage the effects of housing-related injuries and diseases, including, for example, lead poisoning from exposure to lead paint or water pipes; acute respiratory diseases from indoor air pollution; chronic diseases such as asthmas and allergies related to pollution, dampness and mould; and water and sanitation-related diseases.

7.7.4 Promote good occupational health in building trades

A "life cycle" approach should be taken to occupational health practices in the construction industry, considering risks inherent to: initial construction, use, retrofit, and demolition, and also including the design, construction and retrofit of more climate-friendly and energy-efficient buildings.

Building trades should advocate for good occupational health practices, such as the routine use of protective clothing and masks, safe scaffolding and safe equipment. This should also include heightened awareness of construction product hazards in the installation/removal phases, as well as awareness of hazardous products that should never be used at all (e.g. asbestos).

Health actors should work more closely with construction industry and labour stakeholders to support development of policies and capacity for healthy workplaces, including more systematic procedures for occupational risk assessment and management.

Employers should routinely provide information and training for construction workers on how to prevent occupational hazards as well as specialized training in prevention and control measures related to particular construction and retrofit tasks.

Construction workers receive training as part of an energy efficiency and housing upgrade project in Kuyasa, a low-income area of Cape Town, South Africa. The project aims to help reduce illnesses from housing-related heat stress, cold exposure, and other health risks. It is the first in South Africa to qualify for finance under the Clean Development Mechanism of the United Nations Framework Convention on Climate Change (UNFCCC). (Photo: Nic Bothma/Kuyasa CDM)

Notes

Notes